Digital Twin for Gear Wear Monitoring and Prediction

This book presents recent research developments and integrated methodologies for digital twin gear wear monitoring and remaining useful life prediction for rotating machinery. It describes a comprehensive framework for identifying wear mechanisms, developing dynamic gearbox models, and implementing online monitoring schemes that track the evolution of abrasive wear and fatigue pitting. The methodologies introduced allow for accurate assessment of tooth profile changes and surface integrity without requiring operational stoppage. Simulations and dynamic model implementations in this book are constructed using the MATLAB® and Simulink® software packages.

Features:

- Gives a systematic investigation of vibration-based techniques to distinguish between fatigue pitting and abrasive wear.
- Develops an integrated monitoring and prediction framework using a dynamic gear model and wear model.
- Includes a novel digital-twin approach that regularly updates model coefficients using measured vibration data to ensure prediction accuracy.
- Discusses the impact of macro- and micro-level wear on dynamic contact forces and vibration characteristics.
- Provides experimental validation through run-to-failure tests conducted under both dry and lubricated conditions.

This book is aimed at researchers and graduate students in mechanical engineering, signal processing, machine condition monitoring, and reliability engineering.

Digital Twin for Gear Wear Monitoring and Prediction

Ke Feng, Qing Ni, and Hanbin Zhou

CRC Press
Taylor & Francis Group
Boca Raton London New York

CRC Press is an imprint of the
Taylor & Francis Group, an **informa** business

Designed cover image: Ke Feng, Qing Ni, and Hanbin Zhou

First edition published 2027
by CRC Press
2385 NW Executive Center Drive, Suite 320, Boca Raton FL 33431

and by CRC Press
4 Park Square, Milton Park, Abingdon, Oxon, OX14 4RN

CRC Press is an imprint of Taylor & Francis Group, LLC

ISBN: 9781041167242 (hbk)
ISBN: 9781041168638 (pbk)
ISBN: 9781003686736 (ebk)

DOI: 10.1201/9781003686736

Typeset in Times
by Newgen Publishing UK

Dedication

To our families and friends.

Contents

Preface

The field of mechanical engineering is continuously evolving, driven by the need for more efficient, reliable, and predictive maintenance technologies. At the heart of many industrial applications, gear systems play a crucial role in the functionality and performance of machinery. Over time, gear wear poses significant challenges, leading to potential failures and costly downtime. It is within this context that we present '*Digital Twin for Gear Wear Monitoring and Prediction*,' a comprehensive exploration of the innovative intersection between digital twin technology and gear wear monitoring.

Our journey into this subject began with a shared interest in advancing the state of the art in mechanical system diagnostics and prognostics. With the growing complexity of industrial systems, traditional methods of gear wear monitoring are often insufficient. We recognised the potential of digital twin technology – a virtual representation of a physical system that evolves in real time with its physical counterpart – to revolutionise how we monitor and predict gear wear. This book aims to provide both a theoretical foundation and practical methodologies for integrating digital twins into gear wear monitoring processes.

In Chapter 1, we set the stage by discussing the significance of gear wear in mechanical systems and introducing digital twin technology. This chapter outlines our objectives and provides a roadmap for the rest of the book. The subsequent chapters delve deeper into specific aspects of gear wear monitoring and prediction, each building upon the last to form a cohesive guide for researchers, engineers, and practitioners.

Chapter 2 reviews recent advancements in vibration-based gear wear monitoring and prediction techniques, emphasising the critical role of vibrations and digital twins in indicating wear. Chapter 3 presents the core architecture of the digital twin methodology for gear wear monitoring and prediction, providing a guiding framework for the detailed methodologies developed in Chapters 4–7. Chapter 4 focuses on identifying wear mechanisms and tracking wear evolution through vibration analysis, providing valuable insights into the practical applications of our methodologies.

In Chapter 5, we detail the development and validation of dynamic models essential for accurate wear monitoring. This forms the basis for Chapters 6 and 7, where we introduce digital twin-based approaches for monitoring and predicting abrasive wear progression and fatigue pitting, respectively. These chapters highlight the integration of dynamic models and vibration data with digital twins, showcasing the power of this technology in real-world scenarios.

Finally, Chapter 8 concludes our exploration by summarising the key findings and discussing future directions in digital twin technology for gear systems. We also include appendices with a glossary of terms to aid readers in navigating the technical language used throughout the book.

This book is the culmination of collaborative efforts and research conducted at the University of New South Wales, the University of Technology Sydney, the University of British Columbia, Imperial College London, the National University of Singapore, and Xi'an Jiaotong University. We are grateful to our colleagues, supervisors, and research partners for their invaluable contributions. Our hope is that this work will inspire further advancements in gear wear monitoring and serve as a foundational reference for those embarking on similar endeavours.

We invite you, the reader, to explore the exciting potential of digital twin technology in enhancing gear wear monitoring and prediction. Whether you are an academic researcher, a practising engineer, or a student, we believe you will find valuable insights and practical tools within these pages.

Ke Feng
Xi'an Jiaotong University
Qing Ni
Northwestern Polytechnical University
Hanbin Zhou
Xi'an Jiaotong University

Author Biographies

Ke Feng is a Full Professor at Xi'an Jiaotong University, is a Marie Curie Fellow, and is ranked among the 'Stanford/Elsevier Top 2% Scientists.' He earned his bachelor's and master's degrees from the University of Electronic Science and Technology of China and his PhD from the University of New South Wales. He has held positions at renowned institutions such as the University of British Columbia, the National University of Singapore, and Imperial College London. His research areas include digital twins, machine learning, signal processing, fault diagnosis, fatigue, and wear analysis, among others. In 2023, he was awarded the title of 'Emerging Leader' by the Royal Physical Society Journal. He currently serves as an Associate Editor and Editorial Board Member for several international journals, including IEEE Transactions on Industrial Informatics, Information Fusion, IEEE Internet of Things Journal, and Structural Health Monitoring. He has led numerous international collaborative projects, including the Horizon Europe, UKRI projects, the National Natural Science Foundation Excellent Young Scientists Fund, and the key projects under the National Key Research and Development Program of China. He has also received the 'Second Prize of the China Aviation Science and Technology Award' and the 'Second Prize of the Vibration Engineering Society Science and Technology Award.'

Qing Ni is a Professor at the School of Artificial Intelligence, OPtics and ElectroNics, Northwestern Polytechnical University, China. She earned her PhD degree from the University of Technology Sydney, Australia, in 2023. She worked at the University of Technology Sydney, Australia, from 2023 to 2025. Her main research interests include large language models, digital twins, signal processing, prognostics and health management. Currently, she serves as Associate Editor or Editorial Board Member for IEEE Transactions on Industrial Informatics, Engineering Applications of Artificial Intelligence, Neurocomputing, Journal of Intelligent Manufacturing, IEEE Transactions on Instrumentation & Measurement, and IEEE Sensors Journal. She also holds positions as Youth Editor and Guest Editor for several journals, in addition to chairing sessions at international conferences. She is recognised as an Emerging Leader by the Measurement Science and Technology journal and World's Top 2% Scientists by Stanford University in 2024.

Hanbin Zhou is currently a PhD candidate in the School of Mechanical Engineering at Xi'an Jiaotong University, China. He received an M.S. degree in Mechanical Engineering from Central South University in 2025. His current research interests focus on cognitive digital twins, machine condition monitoring, signal processing, and artificial intelligence. He has participated in research projects funded by the National Natural Science Foundation of China (NSFC). He holds four authorised Chinese invention patents. He has authored research papers in reputable journals such as Measurement Science and Technology. He also serves as a reviewer for leading journals, including IEEE Transactions on Industrial Informatics.

1 Introduction

1.1 INTRODUCTION

Gearboxes are essential components in many rotating machines and are characterised by high transmission efficiency, reliable operation, and a precise and constant transmission ratio. Owing to these advantages, gearboxes are widely implemented in various mechanical transmission systems across multiple industries. During operation, particularly under harsh environmental conditions, material degradation may occur, making gearboxes susceptible to wear – a progressive loss of material caused by the contact of mating tooth surfaces under relative motion [1–3]. Wear may induce local stress concentrations, which can act as initiation sites for other modes of gear failure such as spalling, gear root cracking, and gear tooth breakage [4–6]. Therefore, monitoring the evolution of gear wear and scheduling necessary maintenance in advance are essential for preventing catastrophic failures and improving the operational availability of gear systems [7].

Abrasive wear and fatigue pitting are among the most common wear mechanisms observed throughout the service life of gears [8]. Abrasive wear, typically caused by insufficient or contaminated lubrication [9], is characterised by a high material removal rate on the gear surface, which leads to changes in the tooth profile. Such changes in the tooth profile reduce tooth thickness and consequently increase the risk of tooth breakage. In contrast, fatigue pitting arises from repetitive rolling–sliding contact and produces tooth profile degradation at a much slower rate. It is typically identified by the presence of pits on the gear surface [4]. Fatigue pitting may also lead to the formation of surface spalls across the entire tooth width, thereby reducing surface durability and potentially causing tooth breakage. Therefore, identifying abrasive wear and fatigue pitting, as well as monitoring the associated surface degradation processes (e.g. tooth profile changes and pitting propagation), are critical topics within the field of wear analysis. Moreover, monitoring and predicting wear progression enable timely maintenance actions to prevent catastrophic failures, which substantially enhances Prognostics and Health Management (PHM) capabilities. Therefore, the ability to detect and assess wear mechanisms constitutes a crucial component of machine condition monitoring, as briefly outlined below.

Machine condition monitoring includes the detection, diagnosis, and prognosis of an abnormal condition of a machine. Prognostics remains the least developed yet

DOI: 10.1201/9781003686736-1

potentially the most valuable of these three phases. Several major techniques are employed in machine condition monitoring [10], including performance analysis, vibration analysis, lubricant and wear debris analysis, thermography, and acoustic emission analysis. Among these techniques, wear debris analysis and vibration analysis are the most commonly applied methods for gear condition monitoring. In practice, wear particle analysis is widely adopted for monitoring gear wear. Specifically, the concentration, size, and size distribution of wear particles serve as effective indicators of the overall wear condition of a machine. In addition, particle size, shape, and surface morphology provide valuable features for identifying wear mechanisms. However, wear particle analysis is often carried out offline, which can be time-consuming and costly [10, 11]. In contrast, vibration signals capture the instantaneous dynamic response of gears and can be readily acquired online. It is well established that vibration signals encode gear tooth profile information within their deterministic components (e.g. gearmesh harmonics and sidebands) [12, 13]. Recent preliminary studies have also suggested a relationship between gear surface morphology and the random components of vibration signals [14, 15]. Vibration analysis is therefore considered a highly promising approach for efficient real-time gear wear monitoring. Although vibration-based techniques are widely used for gear fault detection, diagnosis, and prognostics, their application to wear monitoring has been limited because wear-related vibration features are highly complex and difficult to extract. To fully exploit the wear-related information contained in vibration signals and to develop vibration-based techniques for applications such as online gear wear monitoring and prediction, substantial further research is required to establish methods that are suitable for practical implementation.

1.2 RESEARCH GOALS

The ultimate goal of this book is to develop vibration-based techniques for wear monitoring and for predicting the remaining useful life (RUL) of gear systems, thereby complementing existing capabilities in fault detection and diagnosis. The specific objectives of this book are as follows:

1. to investigate wear-induced vibration characteristics and to identify wear-related vibration features for distinguishing between abrasive wear and fatigue pitting, which are two common gear wear mechanisms.
2. to develop a vibration-based online method for monitoring gear wear processes using the identified vibration indicators.
3. to develop an integrated framework for monitoring and predicting gear wear propagation by combining computer simulations, empirical wear models, and the characteristic vibration responses of operating machines.

1.3 STRUCTURE OF THIS BOOK

This book begins with a review of the relevant literature (Chapter 2), followed by a description of the methodology developed to achieve the above objectives in Chapter 3. The development of vibration-based techniques for wear mechanism identification and for tracking wear progression (Objectives 1 and 2) is presented in Chapter 4. Objective 3 is addressed in three steps: (i) Establishing a lumped-parameter dynamic model of the gearbox transmission system; (ii) developing a methodology for monitoring and predicting wear-induced tooth profile changes; and (iii) developing a methodology for the simultaneous monitoring and prediction of tooth profile changes and surface pitting propagation. The development of the gear wear monitoring and prediction framework through these three steps is reported in Chapters 5, 6, and 7, respectively. The book concludes with Chapter 8, which presents the conclusions and recommendations.

This book consists of eight chapters. Summaries of the chapters are provided below:

Chapter 1: Introduction

Chapter 2: This chapter reviews the progress of vibration-based gear wear monitoring, including gear wear modes, the relationships between gear wear and dynamic responses of gear systems, and vibration-based techniques (vibration features and models) for gear wear monitoring.

Chapter 3: This chapter describes the methodology used in this research, including the experimental facilities, the data acquisition system, and the procedures used to achieve the research objectives.

Chapter 4: This chapter introduces wear mechanisms and methods for identifying and tracking the evolution of wear using the cyclostationary characteristics of vibration signals. It also presents the effects of different wear mechanisms (abrasive wear and fatigue pitting) on gear systems, specifically tooth profile changes and surface pitting.

Chapter 5: This chapter describes the development of a lumped-parameter dynamic model of the gearbox test rig at the University of New South Wales (UNSW). The technical details and results of dynamic model validation and calibration are presented.

Chapter 6: This chapter presents the developed methodology for monitoring and predicting gear tooth profile changes. A run-to-failure test was conducted using the gearbox test rig at UNSW under dry (unlubricated) conditions, resulting in abrasive wear of the gear teeth.

Chapter 7: This chapter presents the simultaneous monitoring and prediction of surface pitting and gear tooth profile changes. It extends the work in Chapter 6 by considering an additional wear mechanism, namely surface pitting. A run-to-failure test was conducted on the UNSW gearbox test rig under lubricated conditions, resulting in both tooth surface pitting and mild tooth profile changes.

Chapter 8: This chapter summarises the key findings and articulates the novel contributions to knowledge in this field, and provides recommendations for future work.

REFERENCES

[1] A. Ramalho, J.C. Miranda, The relationship between wear and dissipated energy in sliding systems, *Wear*, 260 (2006) 361–367. https://doi.org/10.1016/j.wear.2005.02.121.

[2] K. Mao, Gear tooth contact analysis and its application in the reduction of fatigue wear, *Wear*, 262 (2007) 1281–1288. https://doi.org/10.1016/j.wear.2006.06.019.

[3] H. Wang, C. Zhou, H. Wang, B. Hu, Z. Liu, A novel contact model for rough surfaces using piecewise linear interpolation and its application in gear wear, *Wear*, 476 (2021) 203685. https://doi.org/10.1016/j.wear.2021.203685.

[4] P.J.L. Fernandes, C. McDuling, Surface contact fatigue failures in gears, *Engineering Failure Analysis*, 4 (1997) 99–107. https://doi.org/10.1016/S1350-6307(97)00006-X.

[5] P.K. Singh, S. Siddhartha, A.K. Singh, An investigation on the thermal and wear behavior of polymer based spur gears, *Tribology International*, 118 (2018) 264–272. https://doi.org/10.1016/j.triboint.2017.10.007.

[6] R. Shao, W. Hu, Y. Wang, X. Qi, The fault feature extraction and classification of gear using principal component analysis and kernel principal component analysis based on the wavelet packet transform, *Measurement*, 54 (2014) 118–132. https://doi.org/10.1016/j.measurement.2014.04.016.

[7] K.F. Brethee, D. Zhen, F. Gu, A.D. Ball, Helical gear wear monitoring: Modelling and experimental validation, *Mechanism and Machine Theory*, 117 (2017) 210–229. https://doi.org/10.1016/j.mechmachtheory.2017.07.012.

[8] G.E. Morales-Espejel, P. Rycerz, A. Kadiric, Prediction of micropitting damage in gear teeth contacts considering the concurrent effects of surface fatigue and mild wear, *Wear*, 398–399 (2018) 99–115. https://doi.org/10.1016/j.wear.2017.11.016.

[9] J.A. Williams, A.M. Hyncica, Mechanisms of abrasive wear in lubricated contacts, *Wear*, 152 (1992) 57–74. https://doi.org/10.1016/0043-1648(92)90204-L.

[10] R.B. Randall, *Vibration-based condition monitoring: Industrial, aerospace and automotive applications*, Wiley, 2011.

[11] Z. Peng, T.B. Kirk, Automatic wear-particle classification using neural networks, *Tribology Letters*, 5 (1998) 249–257. https://doi.org/10.1023/A:1019126732337.

[12] G. Liu, R.G. Parker, Dynamic modeling and analysis of tooth profile modification for multimesh gear vibration, *Journal of Mechanical Design*, 130 (12) (2008) 121402. https://doi.org/10.1115/1.2976803.

[13] C. Hu, W.A. Smith, R.B. Randall, Z. Peng, Development of a gear vibration indicator and its application in gear wear monitoring, *Mechanical Systems and Signal Processing*, 76–77 (2016) 319–336. https://doi.org/10.1016/j.ymssp.2016.01.018.

[14] X. Zhang, W.A. Smith, P. Borghesani, Z. Peng, R.B. Randall, *Use of cyclostationarity to detect changes in gear surface roughness using vibration measurements, asset intelligence through integration and interoperability and contemporary vibration engineering technologies*, Springer, Brisbane, Australia, 2019, pp. 763–771.

[15] Y. Yang, W.A. Smith, P. Borghesani, Z. Peng, R.B. Randall, *Detecting changes in gear surface roughness using vibration signals*, in Proceedings of the Acoustics Conference, Hunter Valley, NSW, Australia, 2015, pp. 15–18.

2 Literature Review

2.1 GEAR WEAR

2.1.1 Gear Wear Modes

Surface wear is a common but inevitable phenomenon during the whole service life of a gearbox [1]. When gear pairs mesh, the tooth flanks will be loaded to maintain contact. The motion of the gear tooth surfaces is a combination of rolling motion and sliding motion. The sliding component occurs where the surface velocities of the two contacting teeth are different [2]. The sliding motion can cause material removal from the gear teeth, which results in gear mass reduction, that is, gear wear. Gear wear can be sorted based on the wear mechanisms as follows:

- abrasive wear: Particle contamination or insufficient lubrication, which could lead to sliding contact resulting in abrasive wear. Abrasive wear leaves radial scratches on the gear surface and causes changes to the geometry of gear teeth [3], as shown in Figure 2.1.
- fatigue: Surface fatigue results in the removal of material, and it will leave cavities on the flank of the gear tooth. Normally, surface fatigue includes pitting and case crushing [3]. Case crushing often occurs in heavily loaded, case-hardened gears. Compared with case crushing, fatigue pitting is more common during the service life of gears. Fatigue pitting is caused by cyclic loading conditions, resulting in fatigue cracks either on the surface of the gear tooth or at a shallow depth below it. The initial crack usually propagates for a short distance in a direction roughly parallel to the tooth surface before turning or branching to the surface. When the cracks have grown long enough to separate a piece of the surface material, fatigue pitting is formed [4], as shown in Figure 2.2.
- scuffing: When there is substantial sliding motion between mating teeth under lubrication conditions, the excessive temperature at the sliding asperity contacts may result in the breakdown of all protective lubricating films. As a consequence, the softer asperities can deform plastically and transfer to the mating surface. This is usually accompanied by a rapid increase in wear rate

DOI: 10.1201/9781003686736-2

FIGURE 2.1 Extremely worn gear due to abrasive particles in the lubricant. Photograph by the American Gear Manufacturers Association [4].

or even seizing of the sliding pair [5]. This phenomenon is known as scuffing, as shown in Figure 2.3.

- corrosive wear: Corrosive wear is a visible wear type as a surface deterioration, as shown in Figure 2.4. It is mainly caused by chemical reactions with active ingredients in the lubricant [3]. Corrosive mild wear in gears is usually introduced by lubricant additives intended for preventing scuffing failure, such as extreme pressure additives [5].

In practical applications, abrasive wear and fatigue pitting are the most common wear phenomena in gear systems [6]. Therefore, in this book, the two are chosen as the objective of the study.

2.1.2 Differences Between Abrasive Wear and Fatigue Pitting

Based on the comparison of abrasive wear and fatigue pitting shown in Figure 2.1 and Figure 2.2, there are two major differences between these two surface degradation mechanisms. First, abrasive wear often has a high wear rate and can result in

FIGURE 2.2 Fatigue pitting. Photograph by the American Gear Manufacturers Association [4].

noticeable accumulated material removal in the gear tooth thickness, i.e. changing the gear tooth profile over a certain period. Normally, tooth profile change is measured in millimetres and can be named as macro-level wear, which is illustrated in Figure 2.5. In Figure 2.5, it can be seen that the maximum tooth profile change caused by abrasive wear occurs at the addendum and dedendum of the gear tooth, while the profile change is minimal at the pitch line. Consequently, the worn tooth has a double-scalloped tooth profile. The reason is that the sliding velocity, which is proportional to the wear depth in the Archard wear model, is theoretically zero (and zero wear depth) at the pitch line and reaches the maximum sliding speed (and wear depth) at the addendum and dedendum. This phenomenon has also been observed in gear wear simulation [7] and gear wear tests [4]. In contrast, fatigue pitting has a lower wear rate in the tooth thickness direction, resulting in negligible effects on gear tooth profile geometry, unless it is extremely severe. Therefore, these two wear mechanisms can often be differentiated at different scales: Abrasive wear at the macro scale and fatigue pitting at the micro scale.

Second, in view of gear surface morphology, it can be found that compared with fatigue pitting, abrasive wear tends to produce a surface with a relatively short wavelength in the sliding direction, which results in a high spatial frequency [9]. As for

FIGURE 2.3 Typical scuffing failure in gears. Photograph by B. Bhushan [5].

fatigue pitting, the detachment of material fragments from the gear tooth surface results in localised valleys with long-wavelength which corresponds to a relatively low spatial frequency.

The differences in the features/characteristics of abrasive wear and fatigue pitting are summarised in Table 2.1.

2.2 GEAR WEAR EFFECTS ON THE VIBRATIONS OF GEAR SYSTEMS

There is a two-way relationship between the wear process and gear dynamic characteristics. In general, gear wear can result in the alteration of gear tooth profile geometry or reduction of contact area, which will change the geometric transmission error (GTE) and meshing stiffness of the gear system, and then the dynamic characteristics will be affected, including the dynamic contact force and its distribution. As a consequence, the level of vibration and harmful noise tend to increase [10]. In turn, the change of dynamic contact force could alter and accelerate the gear wear process. The two-way relationship between gear wear and gear dynamics will produce complex gear dynamic responses and vibration features, which brings huge

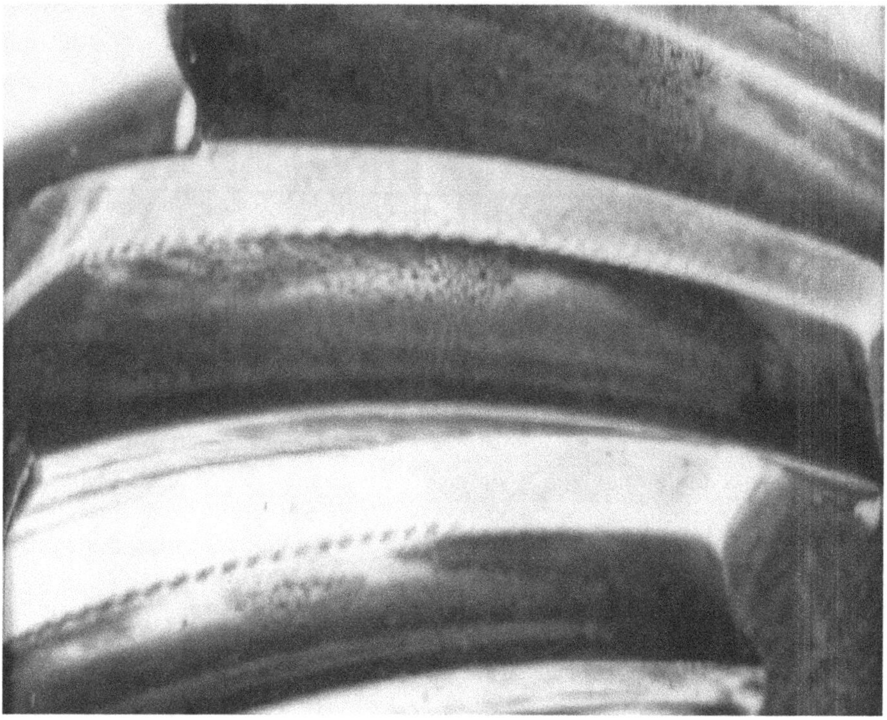

FIGURE 2.4 Corrosion damage. Photograph by H. Endo [3].

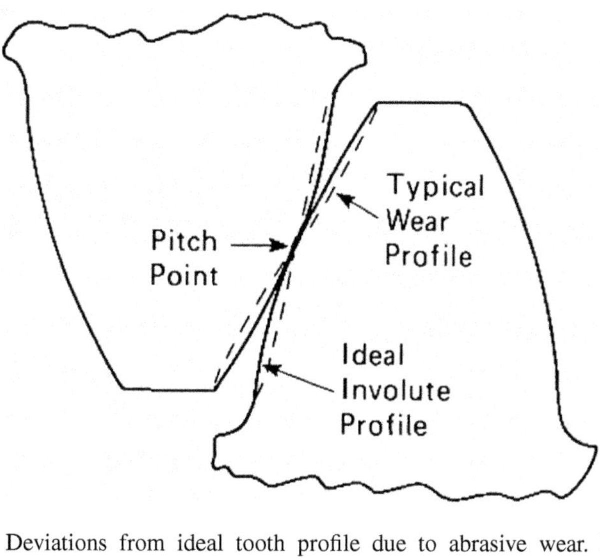

FIGURE 2.5 Deviations from ideal tooth profile due to abrasive wear. Made by R.B. Randall [8].

TABLE 2.1
Differences Between Abrasive Wear and Fatigue Pitting

Wear type	Wear rate	Morphology (spatial frequency)	Final form
Abrasive wear	High	High	Tooth profile change (together with a rough gear surface)
Fatigue pitting	Low	Low	Valleys on the certain region of gear tooth (between gear root and pitch line)

challenges to condition monitoring on gear wear progression compared with other failures, such as gear tooth root crack, tooth surface spalling, and tooth breakage.

With consideration of the contact patterns, abrasive wear and fatigue pitting have different impacts on the vibrations of the gear system at two different scales. Existing studies and understanding of gear wear effects on vibrations are presented in the macro-level and micro-level below.

Macro-level wear (gear tooth profile change) is a kind of geometric deviation from the ideal gear tooth profile, which is serviced as a GTE to the gear system. The macro-level wear can lead to an increase in the strength of gear meshing harmonics in the vibration signal [11, 12]. Meanwhile, due to the wear-induced gear tooth profile change, the load distribution on the tooth surface will also be altered, and therefore, the dynamic characteristics of mating gears will be changed. The effect of macro-level wear on vibrations can be illustrated in Figure 2.6.

As explained in Section 2.1, abrasive wear can easily result in the gear tooth profile change, while fatigue pitting usually does not modify the gear tooth profile if gears are lubricated with a relatively low wear rate. However, both abrasive wear and fatigue pitting have significant impacts on the micro-geometry of the gear tooth surface. These two wear mechanisms can induce different surface morphologies at the micro-level. Abrasive wear can lead to the creation of protrusions (i.e. lumps) distributed from gear root to tip uniformly, while fatigue pitting induces the occurrences of valleys on the gear surface, normally distributed from the gear root to pitch line.

The micro-level wear generates a rough gear surface, which increases the friction force between the mating gears and raises the overall vibration level and its frequency characteristics [13]. The energy induced by micro-level wear might be very low in comparison with the macro-geometry of the gear surface. The micro-level wear would induce random vibrations, namely sliding vibrations [14], so it won't be represented in the gear meshing harmonics (deterministic signals); thus, it is not easy to distinguish and extract micro-level wear information from other effects in the original measured vibration signals [15].

The effects of abrasive wear and fatigue pitting, at both macro-scale and micro-scale, on vibrations are summarised in Table 2.2.

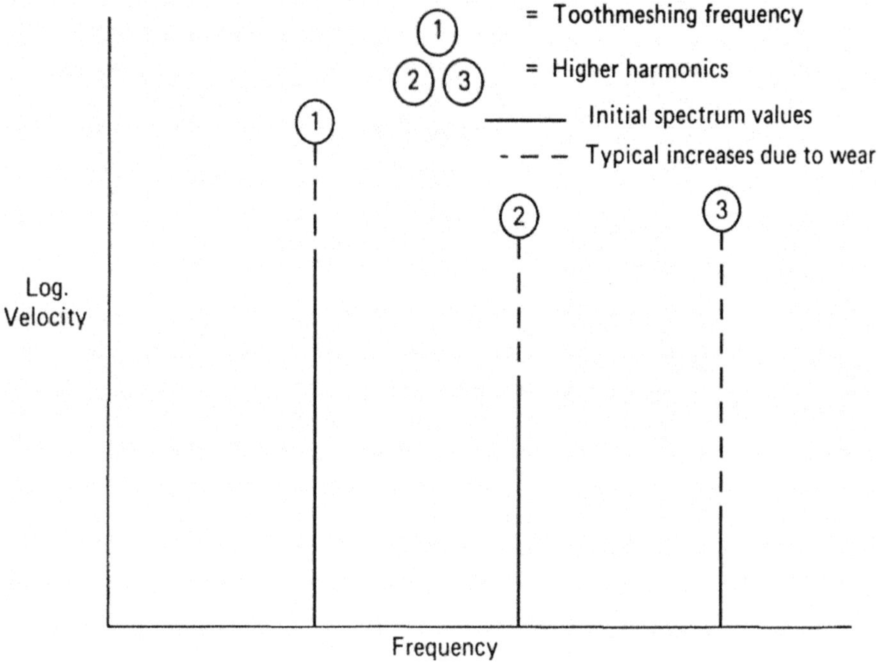

FIGURE 2.6 Typical vibration spectrum due to wear. Made by R.B. Randall [8].

TABLE 2.2
Effects of Wear on Vibrations [8, 14, 16, 17]

Wear type	Macro-level	Micro-level
	Gear meshing harmonics (deterministic components) change	Sliding-induced vibration (random components) change Magnitude
Abrasive wear	Significant	Increase
Fatigue pitting	Slight	Increase

From Table 2.2, it can be found that these two wear mechanisms have distinct impacts on different vibration features. In practice, when fatigue pitting propagates, abrasive wear may also co-exist due to oil contamination. The abrasive wear could help remove high asperities and then lead to a smoother gear surface and improved lubrication, which can help prevent the occurrence of fatigue pitting. Conversely, the occurrence of fatigue pitting can break the oil film and lead to a contact pressure concentration, which could promote the abrasive wear process [6]. This is a coupling effect between abrasive wear and fatigue pitting, which results in complex vibration characteristics and makes it difficult to extract wear-related vibration features and develop specific vibration-based indicator(s) for wear mechanism identification and evolution tracking. Therefore, the vibration-based techniques for gear wear monitoring are rather rare. In the following, the existing vibration-based gear wear monitoring methodology, using vibration features and models, will be reviewed and summarised.

2.3 USE OF VIBRATION FEATURES FOR GEAR WEAR MONITORING

Current wear monitoring work (using vibration features) mainly focuses on gear wear evolution tracking [8, 14, 16–24]. Moreover, most of this research aims to monitor gear tooth profile change (at the macro-level) [8, 17–23]. In contrast, only a handful of studies are designed for micro-level wear monitoring, such as detecting surface roughness changes [14, 16] or monitoring fatigue pitting propagation [24]. Compared with wear evolution tracking techniques, up to now, the vibration-based techniques for wear mechanisms identification are rarer. Therefore, in the following, existing vibration-based researches for wear evolution tracking are presented first. Then, inspired by these researches, the potential researches for wear mechanism identification will be discussed and summarised.

2.3.1 Vibration Feature-Based Wear Evolution Tracking

As discussed in Section 2.2, abrasive wear (or extremely severe fatigue pitting) could lead to gear tooth profile change (macro-level wear) with an increase in the overall energy of the vibration signal and the magnitude of gear meshing harmonics. Therefore, the relationship between signal energy or gear meshing harmonics and gear wear severity was investigated.

Root mean square (RMS) (Eq. (2.1)) is widely used for reflecting the amplitude (or power) of the vibration signal. Considering the worn gear would bring in geometric deviation from the ideal gear tooth involute and then result in a stronger vibration, there are some researchers in references [19, 20] using RMS to monitor the gear wear process. It has been found that the RMS value has a positive relationship with the gear wear severity. In addition, to improve the sensitivity and reliability of RMS for detecting gear wear change, a sample parameter, named matched filtered RMS, was reported in reference [21]. This parameter was defined to be the logarithmic value (expressed in dB) of the averaged power ratio between components of the current vibration signal and those of the reference signal. Compared with classical parameters such as RMS and peak values, it is easy to trend and performs better in tracking the gear wear process [18].

$$\mathrm{RMS}_{\mathrm{x}} = \sqrt{\frac{1}{N}\left[\sum_{i=1}^{N}(x_i)^2\right]} \tag{2.1}$$

As an extended version of RMS, an indicator named energy ratio (ER) was proposed in reference [17]. It was defined in Eq. (2.2) as the difference signal d divided by the RMS of the signal containing only the regular meshing components y_d.

$$ER = \frac{RMS_d}{RMS_{y_d}} \quad (2.2)$$

ER increases with wear severity when wear occurs uniformly on the tooth surface, as in this case RMS_d would be expected to increase while RMS_{y_d} would decrease.

However, RMS and its extended versions mainly focus on the changes in signal power, and thus may not be able to reflect the changes in the signal spectral distribution, which is also closely related to gear wear. Therefore, some studies have started to focus on the signal spectral distribution change due to the gear wear processes.

With consideration of the wear pattern, the effects of uniform wear on gear mesh harmonics were investigated in reference [8]. In reference [8], the author stated that systematic wear would tend to produce a kind of profile deviation which is indicated in exaggerated form in Figure 2.5. Consequently, the higher-order harmonics of the tooth meshing will increase. Therefore, the amplitude of higher-order meshing harmonics could be a reliable way of detecting uniform wear at its early stage. With this knowledge as a basis, the first three gear mesh harmonics of the spectrum and quefrencies of the cepstrum were used in reference [22] to monitor the gear wear process.

However, during the gear wear process, the average gear tooth working profile/surface will steadily deviate further from the ideal involute geometry profile, but the changes in the gear tooth meshing harmonics are not determined. That is, all the gear meshing harmonics could vary differently, and each gear meshing harmonic may increase in a certain period but decrease in the following period. Due to this complex situation, using only one or several specific meshing harmonics may not be sufficient to monitor the gear wear process. Therefore, all gear meshing harmonics with significant energy were taken into consideration in reference [18]. The sideband ratio (SBR) proposed in reference [25] was then extended and modified into two new indicators: The averaged logarithmic ratio (ALR) and moving averaged logarithmic ratio (mALR). ALR can be used to reflect the wear effects on the gear degradation state, while mALR shows immediate changes in the gear degradation state within each short time interval. The performance of these two developed indicators was evaluated by two sets of tests with different initial gear tooth surfaces.

In theory, surface wear will cause a gradual change in the mechanical properties of the gear transmission system (most notably in the gear tooth geometry profile and the gear meshing stiffness), and therefore, a gradual change in the gear contact mechanism that generates the vibration signal. Thus, the difference between the vibrations of healthy gears and those of worn gears can be used to represent the gear wear propagation process.

In reference [23], an indicator named Model Prediction Error (MPE) was used to track the gear wear process (tooth geometry profile change). An auto-regressive (AR) model was used to predict the current state of the vibration signal based on the historical data, and the 'prediction error' is the difference between the predicted signal and the current measured signal, which can be used to indicate the gear wear process. A comparison with classical indicators was made, such as the FM0, FM4, NA4, and RMS. It proves that MPE has a better performance than those indicators in gear wear propagation monitoring.

Compared with the above-mentioned researches for macro-level wear severity assessment, researches for micro-level wear monitoring are quite rare. The reason is that the micro-level wear-induced vibration is a random vibration with low energy, and it is not easy to distinguish micro-level wear information from other effects in the original measured vibration. In the following, existing researches for micro-level wear assessment will be introduced.

Recent developments [14, 16] have shown that the information on gear tooth surface roughness (induced by abrasive wear or fatigue pitting) can be detected using a cyclostationary-based approach. The cyclostationarity of vibration is caused by

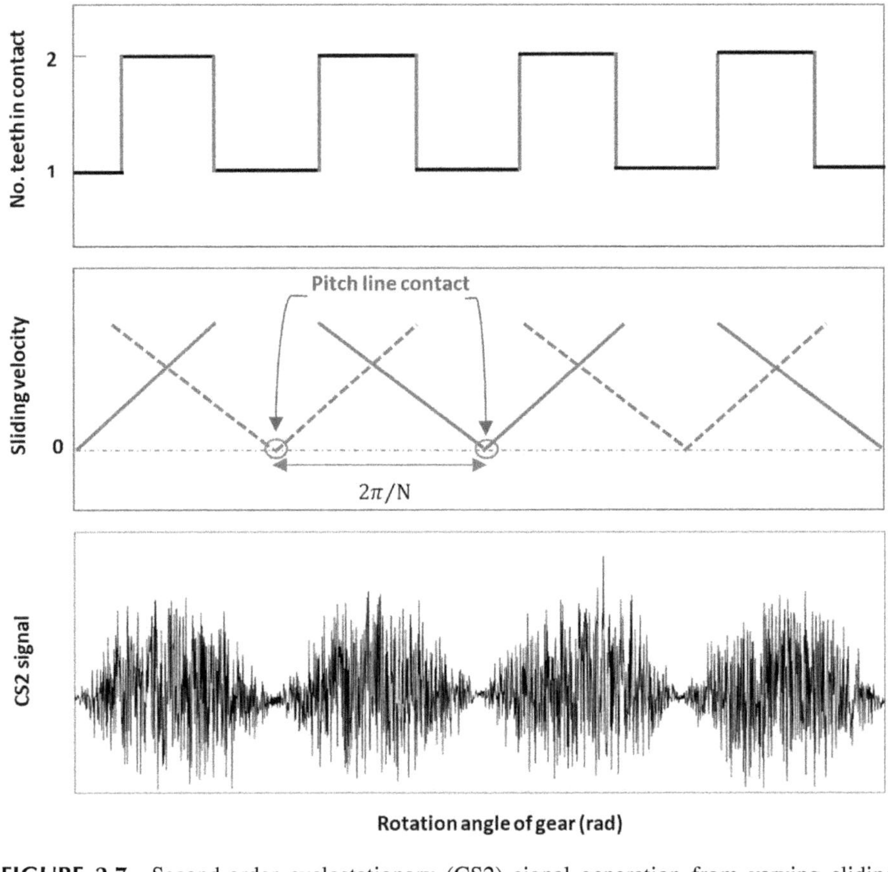

FIGURE 2.7 Second-order cyclostationary (CS2) signal generation from varying sliding velocity in mating gears. Made by Y. Yang [16].

the unique kinematic characteristics of gear transmission systems, as depicted in Figure 2.7. In the study [16], an indicator to measure the degree of second-order cyclostationarity (namely ICS2), proposed in reference [26], was used to monitor the gear surface roughness change. In the experimental part, an approximate positive correlation was found between the degree of second-order cyclostationary indicator (ICS2) of vibration and surface roughness. However, the relationship was found to be more complex by a later investigation [14], based on a wider range of roughness values and a longer experimental duration. To date, insufficient conclusions can be drawn.

In reference [24], the authors assumed that the strength of amplitude and frequency modulation is correlated with the gear wear severity, and used a correlation coefficient to quantify the difference between the reference signal (vibrations of healthy gears) and the current measured vibrations, then linked it to gear wear severity. The difference from the work conducted in reference [23] is that the correlation coefficient used in reference [24] is for monitoring fatigue pitting propagation (micro-level wear), whose information is hard to detect in the deterministic part of the vibration signal. Therefore, a residual signal, after removing gear meshing and shaft harmonics, was used for correlation coefficient calculation. Five natural pitting propagation tests were conducted in the laboratory to verify the effectiveness of this approach.

Based on the above literature review, the existing vibration feature-based techniques for gear wear evolution monitoring can be summarised in Table 2.3. From Table 2.3, it can be found that the vibration feature-based techniques for gear monitoring are

TABLE 2.3
Studies of Vibration Feature-Based Gear Wear Evolution Tracking

References	Main techniques	Purpose
[17–21]	RMS, ER	Accumulated gear wear evolution tracking (at macro-level)
[8, 22]	Gear mesh harmonics or quefrencies	Accumulated gear wear evolution tracking (at macro-level)
[18, 22]	Sideband energy ratio, sidebands	Accumulated gear wear evolution tracking (at macro-level)
[23]	Auto-regressive model, then prediction error	Accumulated gear wear evolution tracking (at macro-level)
[24]	Correlation coefficient-based approach	Fatigue pitting propagation monitoring (at micro-level)
[14, 16]	ICS2 to monitor roughness change	Gear surface roughness monitoring (at micro-level)

quite limited and general. Most of the studies focus on tracking macro-level wear progression (gear tooth profile change), which can be easily detected and monitored in the deterministic components of vibrations. In contrast, studies for micro-level wear, such as fatigue pitting or abrasive wear-induced surface roughness change, are rather rare. The reason is that the vibration characteristics of micro-level wear are weak, and most of them are contained in the random components of vibrations, which are easily masked by unrelated signals from other vibration-generating mechanisms or white noise. This challenge brings huge difficulties to extracting useful vibration features and monitoring micro-level gear wear evolution. Therefore, researches on investigating the internal relations between micro-level gear wear and measured vibration features, which could support the development of vibration-based fatigue pitting propagation, are in vital need and of great importance.

2.3.2 Vibration Feature-Based Wear Mechanism Identification

To date, there has been very limited work on vibration-based wear mechanism identification, and the existing approaches for wear mechanism identification mainly rely on visual inspection of a worn surface and/or its wear particles generated from the surface. Based on the literature review, there are several researches suggesting their potentials for wear mechanism identification, which will be presented as follows.

A phenomenon was found in reference [27], that is, fatigue pitting information is contained in the low-frequency range of vibrations. In reference [27], artificial pits were introduced to all teeth of the pinion with different sizes to simulate different pitting severities, then the mean frequency variation of a scalogram was used to detect the pitting damage. The experimental results showed that the mean frequency decreased when the severity of pitting increased. This suggested that pitting has effects on the low-frequency part of vibrations.

Even though insufficient conclusions can be drawn in references [14, 16], these two suggested that the surface morphology information could be detected in sliding-induced vibration. Reference [14] also mentioned that the wavelength of surface asperities might have an impact on the surface roughness monitoring results. This information can be used to separate abrasive wear and fatigue pitting.

Based on the findings in references [14, 16, 27, 28], the carrier frequency information of sliding-induced vibrations might have the potential to identify abrasive wear and fatigue pitting.

From the above literature review, it can be found that the unique surface features induced by different gear wear mechanisms have not been well explored and investigated, which restricts the potential of vibration-based techniques for gear wear mechanism identification. Therefore, vibration feature-based techniques for gear wear monitoring with consideration of special vibration characteristics induced by different gear wear mechanisms are needed.

2.4 MODEL-BASED GEAR WEAR MONITORING TECHNIQUES

Gear wear simulation has significant benefits for gear wear monitoring and prediction. The gear wear simulation mainly uses the gear meshing mechanism, gear wear mechanism, and vibration characteristics to establish dynamic models, tribological (wear) models, and their interactions. Then, with the help of the established models, responses under different health conditions can be simulated and evaluated, and fault symptoms can be disclosed and concluded for fault diagnostics and prognostics [29–31]. In the gear wear simulation, gear dynamic models are concerned with the relationship between dynamic properties (stiffness, transmission error, friction, etc.) and system responses such as vibration responses (dynamic contact forces and vibration signals). Tribological models rely on wear mechanism theory or experimental data to establish a damage propagation model, in which contact pressure distribution, oil film thickness, and/or wear rate are studied based on certain inputs including the load, lubricant viscosity, sliding velocity, and surface roughness. In the following sections, the research progress on gear wear model developments will be presented.

2.4.1 Dynamic Models of Spur Gearboxes

Contact force is an important input for the tribological model [32–34]. By calculating the contact force, the contact pressure can then be calculated based on Hertzian contact theory [35, 36], and the gear wear propagation behaviours can be simulated using tribological (wear) models. There are a lot of researches using empirical equations or finite element models to evaluate the gear contact force and its distribution between meshing gear pairs [37–42]; however, most of these methods are effective only under quasi-static conditions. Without considering the inertia effects due to the dynamics of the gear system, the quasi-static contact force can be easily simulated using empirical equations and finite element models. However, in industrial practices, the gear transmission system is usually operated under dynamic operating conditions, and the corresponding responses (e.g., contact force) are quite different from those under quasi-static conditions. Normally, because of inertia effects, the dynamic meshing forces are typically larger than the corresponding quasi-static forces, and their magnitudes and waveforms are quite different [7]. Therefore, the dynamic contact force and its distribution should be properly evaluated to guarantee (i) gear wear propagation behaviours can be simulated, and (ii) accurate wear-induced dynamic responses can be exhibited.

To obtain proper/accurate dynamic contact force during the gear wear process, a dynamic model of the gear transmission system, which can also generate wear-induced dynamic responses (such as vibrations) for wear analysis, is required. In general, the dynamic model includes many parameters. Figure 2.8 shows a typical gear dynamic model, reproduced from reference [43]. In the dynamic model, there are two main types of excitations to generate vibrations of the gear pair. The external excitations are the fluctuations of the applied load and input operating speed, while the internal excitations are generated from time-varying cyclic meshing stiffness, $k(t)$, and GTE, $e(t)$ [44]. The occurrence of gear tooth surface wear can affect the GTE (part of internal excitations) significantly, then correspondingly the responses will change, which can represent different fault symptoms with different wear severities.

Generally, when gear wear occurs, the contact patterns between mating gear pairs can be substantially modified, that is, tooth profile change (induced by abrasive wear)

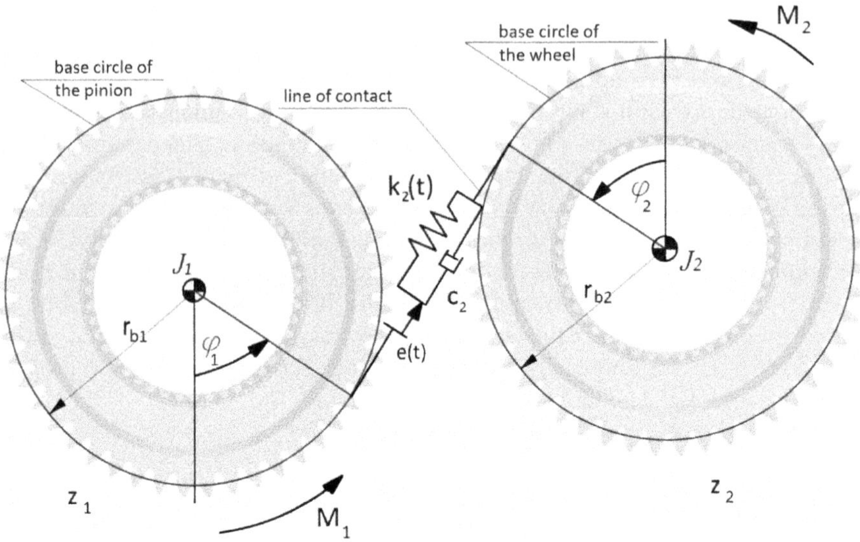

FIGURE 2.8 Dynamic model of a spur gearbox. Made by Y. Temis [43].

and contact area reduction (induced by fatigue pitting). Tooth profile change is one kind of GTE. Both tooth profile change and contact area reduction can alter the gear meshing stiffness. However, considering the scale of gear wear, the meshing stiffness change induced by gear wear can be neglected as it is significantly less important than the transmission error effect [11]. Therefore, only the researches on wear-induced transmission error of the dynamic model will be reviewed.

Owing to the wear-induced GTE, the dynamic load and its distribution between the meshing gear pairs will be altered, which could lead to a dynamic transmission error (DTE) and thus result in changes in vibration and noise levels. Different from the other parameters in the dynamic model, such as backlash, manufacturing error and tooth relief, gear wear will cause a tooth profile change with a certain distribution, which is almost zero around the pitch line and has a maximum value at the root or tip of the gear tooth generally [45]. Different tooth profile changes will cause different corresponding dynamic characteristics and responses. Therefore, to acquire accurate wear-induced dynamic characteristics and responses, GTE should be properly obtained or simulated according to the characteristics of wear caused by tooth profile changes. There are two possible approaches to obtain wear-induced GTE: One is the simulation-based method, and the other one is the experimental method. In the following, researches involving GTE study for gear wear analysis will be introduced.

Experimentally, GTE can be measured using a specialised device. For example, researchers in reference [46] used a designed gear coordinate measurement machine to obtain a large number of lead traces, each containing around 200 measurement points that are aligned using a single profile trace to obtain a three-dimensional measurement of the actual gear tooth surface. This approach can acquire wear-induced

GTE accurately. However, when measuring and evaluating the tooth surface changes, the gearbox should be stopped and dismantled, which may bring other failure modes into the gearbox, such as shaft misalignment. Therefore, many researchers chose the simulation-based approach. Since wear-induced tooth profile change is complex, it is not easy to use simple equations such as sine/cosine functions to represent it accurately. Considering that, some researchers [7, 47–49] used the tribological (wear) models to obtain the wear-induced tooth profile changes. This approach can be regarded as an integration of dynamic models and tribological models, and it will be introduced in Section 2.4.3. In the following, tribological models to simulate wear propagation behaviours will be presented first.

2.4.2 Tribological (Wear) Models for Monitoring Wear Depth and Pitting Density

There are different approaches to establish tribological models with consideration of wear modes (abrasive wear, scoring, corrosive wear, etc.). Since abrasive wear and fatigue pitting are the objectives of this book, tribological models of these two wear mechanisms will be reviewed in this section.

As for abrasive wear, the Archard wear model [50] is the most accepted and widely used. The theoretical basis of the Archard wear model is the Archard wear equation, given by:

$$h = \int K_{\text{wear}} P v dt \tag{2.3}$$

where h denotes the wear depth on the gear tooth, v is the sliding velocity of mating gear pairs at time t, P represents the contact pressure, and K_{wear} is a dimensional wear coefficient. The values for v and P can be determined by the parameters of the gearbox and dynamic model. In contrast, K_{wear} will be different under different lubrication conditions. Therefore, the wear coefficient K_{wear} is a major unknown factor, and it is usually determined from experiments [51] or by an approximate wear coefficient model, which is based on the effect of oil film thickness and gear surface roughness [52]. It is extremely difficult to measure the wear coefficient directly from experiments [53]. Consequently, the widely used approach is to evaluate and determine the wear coefficient using empirical models/equations.

In the determination of wear coefficient using empirical models/equations, lubrication plays an important role, and its effect is considered based on the oil film thickness-to-roughness amplitude ratio defined as $\lambda = h_{min} / R$. The minimum oil film thickness h_{min} can be determined by empirical equations with consideration of thermal effects [54], while $R = \left(R_1 + R_2\right) / 2$, where R_1 and R_2 are the RMS values of surface roughness on the pinion and gear [55]. Based on the value of the calculated λ, three lubrication regimes are considered in the simulations to represent the level of interaction between the mating gear surfaces, and the wear coefficient K_{wear} is calculated as follows:

a) if $\lambda > 4$, it means the oil film thickness is sufficient to avoid direct contact between the gear tooth surfaces, wear is neglected, and K_{wear} is set to be zero.

b) if $\lambda \leq 0.5$, it indicates a strong interaction, wear is maximum, and K_{wear} is usually determined based on experimental results.
c) in the intermediate zone, K_{wear} is supposed to be estimated by linear interpolations based on λ.

In conclusion, the relationship between the λ and K_{wear} can be summarised in Eq. (2.4) [7], given by:

$$K_{\text{wear}} = \begin{cases} k_{0,} & \lambda < \frac{1}{2} \\ \frac{2}{7}k_0(4-\lambda), & \frac{1}{2} < \lambda < 4 \\ 0, & \lambda > 4 \end{cases} \tag{2.4}$$

It can be seen that the values of K_{wear} depend on the oil film thickness and surface roughness, which are used to determine λ. Note that k_0 is an initial value of the wear coefficient.

In a gearbox, the interacting gear teeth are always rolling and sliding against each other under high contact pressure, which means that the lubrication state is most likely in the mixed or boundary regime [56]. Therefore, tribological models used to simulate abrasive wear behaviours are almost always developed under boundary lubrication or mixed elastohydrodynamic lubrication (EHL). In the following paragraph, researches on abrasive wear models under boundary lubrication or mixed EHL will be briefly introduced.

As introduced in Eq. (2.4), surface roughness is an important factor in determining the empirical wear coefficient K_{wear}. However, the initial tribological models for abrasive wear were built without consideration of surface roughness update, in other words, the wear coefficient K_{wear} was treated as a constant value during the whole abrasive wear propagation process [46, 47, 57–61]. In these researches, to achieve a gear wear profile close to the real worn gear, authors in references [46, 47] used a comprehensive finite element model to calculate the contact pressure between meshing gear pairs. However, the wear coefficient K_{wear} had not been updated based on Eq. (2.4), meaning that surface roughness remains to be a fixed value without updating, which is not true during the real gear wear process. The issue of surface roughness update was addressed in the work conducted as part of references [62–64]. Time-varying contact parameters (the normal load, radii of curvature, surface velocities, and slide-to-roll ratio) and wear coefficient K_{wear} updating based on surface roughness were considered in a proposed transient mixed EHL model [62]. Then, the transient behaviour of this model was studied, and a fatigue model for a spur gear in combination with the dynamic model was proposed to study the wear-induced characteristics. The model proposed in reference [62] was subsequently employed in references [63, 64] to establish a fatigue model, but for tooth crack failure rather than fatigue pitting. With consideration of surface roughness update during the wear process, a more accurate wear assessment result can be achieved compared with the model without updating surface roughness.

From the above literature review, it can be found that the wear coefficient K_{wear} in most existing tribological models is an empirical value, even with consideration of surface roughness updating during the gear wear process. However, in actual practice, besides the surface roughness, lots of other factors can also affect wear coefficient K_{wear}, such as contamination of the lubricant, operation condition change, and surface morphology change. Therefore, to accurately simulate wear propagation behaviours, it is necessary to obtain the real accurate wear coefficient K_{wear} based on actual measurements using efficient and reliable tools.

Compared with abrasive wear, studies on simulating surface pitting propagation behaviours are sparser, although there are plenty of publications focusing on explaining the process of surface pitting initiation [63, 65–67]. In reference [68], a multi-axial fatigue criterion and an EHL model [62] were combined to develop a fatigue pitting model. Using this model, the progression of micro-pits on the tooth surface was simulated. Similarly, with the help of the fatigue formula and EHL model, simulation of fatigue pitting propagation behaviours under mixed EHL conditions was achieved in reference [6]. Different from reference [68], the competitive behaviours between fatigue pitting and abrasive wear-induced mild wear were also investigated. Both references [6] and [68] involved the EHL model, which is time-consuming due to its high complexity and high-level expert knowledge required for model establishment. It will bring huge challenges to applications in industrial practices. Therefore, it is vital to develop more efficient models/tools to simulate fatigue pitting behaviours. To address this issue, based on the Lundberg-Palmgren model [69], a modified fatigue model with high computational efficiency was proposed in reference [70], and the results of test rig trials and material analyses were presented to demonstrate the effectiveness of the proposed fatigue pitting model.

2.4.3 Integration of Dynamic and Tribological Models for Gear Wear Monitoring

As mentioned in Section 2.4.1, GTE is a key parameter of the gear dynamic model for gear wear analysis. However, it is challenging to acquire an accurate tooth wear profile purely based on experimental or simple analytical approaches. The tribological (wear) model can be used to generate the wear curve on the tooth flank, and then the generated wear curve can be incorporated into the gear dynamic model to simulate vibrations induced by gear wear. This integration of dynamic and tribological models can help reveal the connection between gear wear and vibration characteristics, which can bring significant benefits to gear wear monitoring.

However, up to this point, there are only limited references [7, 48, 49] using the integration of tribological and dynamic models for gear wear monitoring. Among them, authors in references [7, 49] combined the tribological model and dynamic model together to study the coupling effects between surface wear and gear dynamics (such as meshing stiffness, contact force, and vibrations). Then this approach was extended to the planetary gearbox in reference [48]. The rRRR Reference [7] employed a torsional model with a single degree of freedom and then integrated it with a wear prediction model [46] to investigate the interactions between the tooth surface wear and the dynamic characteristics of spur gear system. However, an accurate prediction for gear dynamics relies on a comprehensive dynamic model that can simulate the behaviours of the actual running rig. Reference [7] only included the torsional deflections in gear-shaft systems; the translational effects coming from the shaft bending and bearing radial deflections were not considered, which may degenerate the accuracy of the wear analysis. To solve this problem, later, a 4-degree-of-freedom model including translational motions of gears was introduced in reference [49], and a new dynamic wear analysis method was proposed to study the interactions between tooth surface wear and gear dynamics. However, in reference [49], the authors used simple sine/cosine functions to represent the meshing stiffness and GTE, which differ significantly from the actual application. An inaccurate evaluation of meshing stiffness and GTE in the gear dynamic model will also cause degeneration in the accuracy of the wear analysis.

Based on the discussion of the research publications in the previous sections, it can be seen that studies on the interaction between tribological and dynamic models for assessing gear wear processes are still needed, with consideration of a comprehensive dynamic model together with a proper evaluation of parameters (stiffness and GTE). Therefore, the establishment of a comprehensive dynamic model with proper meshing stiffness and GTE is necessary, which could simulate realistic wear-induced vibrations (compared with the actual running rig) for further wear analysis.

2.5 WEAR PREDICTION TECHNIQUES

The capability to predict the gear wear process would bring enormous benefits in terms of cost and safety to a wide range of industries. In the following, existing studies of vibration-based gear wear prediction techniques will be reviewed.

2.5.1 Prediction of Tooth Profile Change from Abrasive Wear

In reference [71], the wear distribution on the gear tooth was predicted using the Archard wear model for an unlubricated system. Their observations from experiments validated the prediction results, that is, the maximum wear occurs in the dedendum and addendum regions of gears. However, a gear system usually operates under substantial lubrication, as it reduces wear on the gear teeth, lowers noise and vibration, and improves the power conversion efficiency as less energy is irrecoverably lost to wear mechanisms. To address this issue, an EHL model was applied in reference [72] to simulate the wear propagation behaviours and predict the accumulated wear depth under lubrication conditions. But the applied EHL model in reference [72] is time-consuming and requires a high-expert knowledge for establishment. To reduce the computational cost of the EHL model, a simplified EHL model was developed in reference [73], where temperature effects were considered, and the predicted results of gear wear were validated against those obtained by isothermal formulas defined in reference [74]. However, in references [71–73], the contact force was calculated using empirical equations without consideration of the real worn tooth profile geometry, which could degrade the accuracy of prediction results.

Several studies [75–77] used an empirical formula to estimate the contact force between gears, which was then used to estimate the worn tooth profile. The accuracy of these estimates can be improved by applying the finite element method to determine the contact force. A gear surface wear prediction methodology for spur and helical gears was proposed in reference [46]. In this methodology, a finite element model of the gear contact mechanics, in conjunction with Archard's wear equation, was employed to predict gear wear. To ensure the accuracy of prediction results, a specialised measurement device was used to acquire the real worn tooth profile during the gear wear process. This measured worn tooth profile was set as an input to the finite element model to predict the wear propagation progression, and the prediction results were validated through comparison with experimental ones. However, in the approach proposed in reference [46], the gearbox should be stopped and dismantled when measuring the worn tooth profile. The stoppage and subsequent dismantling could introduce other failure modes or affect the progression of existing ones. Researchers in references [78–80] have also employed the finite element method to estimate the contact force between gears, which was used as an input to the Archard wear model, and subsequently used to predict the wear depth of spur and planetary gears. To obtain a more accurate estimation of the contact force and reduce the computational cost, efforts to improve the finite element model were made in references [81, 82]. The results showed that the wear predictions were enhanced with the help of the improved finite element model. However, there is a drawback to using a finite element model, in that it is difficult to properly represent the dynamic characteristics induced by inertia. With the finite element models employed in references [81, 82], the simulated contact force is under quasi-static conditions. However, in actual practice, the gearbox operates under dynamic conditions, and the dynamic contact force is different from the quasi-static one in both magnitude and waveform [7]. To include the dynamic effects into the finite element model, the boundary conditions and mesh generation should be well defined, both of which require high-level expert knowledge, along with increased computational costs. Therefore, the use of a simple finite element model could bring noticeable errors to wear prediction, unless the worn tooth profile can be regularly corrected using actual measurement, as in reference [78]. Thus, a dynamic model is needed to provide the dynamic contact force, which is closest to the actual running test rig. Note that the finite element model mentioned in this book refers in particular to the simple version without well-defined boundary conditions and mesh generations.

In addition to physical model-based gear wear prediction, there are some other approaches proposed to monitor and predict the change of the gear tooth profile due to wear. For example, an integrated prognostic method was proposed in reference [10] for wear prediction in terms of wear depth change. In this hybrid approach, the Archard wear model was used to simulate wear behaviours, and the Bayesian update process was implemented to determine the wear coefficient during the wear process. The predicted results were compared with experimental results from tests using a planetary gearbox that was run to failure. Compared with the wear prediction methods purely relying on physical models or experiments, the results suggested that the integration of the wear model and actual measurements could achieve more reliable and accurate wear predictions. The relationship between gear hobbing processing technique and gear geometric deviation was modelled by applying the improved Particle Swarm Optimisation (PSO) and Back Propagation (BP) algorithms to determine the optimal model parameters in reference [83]. The accuracy of both algorithms was evaluated by the RMS Error between the predicted and experimental values. The results showed that the gear geometric deviations were well predicted and in reasonably good agreement with the experimental data. A statistical model with statistical parameters was proposed in reference [84] to monitor and predict the gear behaviours with extreme tooth profile alteration induced by abrasive wear. The effects of the SiC concentration, applied load, and sliding distance were statistically and physically analysed in detail. Besides, a fusion of ultra-complete independent component analysis (UICA) and parameter estimation (PE) was developed in reference [85] to monitor and predict the severity of gear wear. Even though promising prediction results were achieved in references [83–85], these statistical model-based approaches could not reveal the wear behaviours and the change of gear dynamic responses during the gear wear process, both of which have significant benefits to the understanding of wear mechanism and its impact on the gear system. Moreover, the statistical model-based approach heavily relies on a huge amount of experimental data, which limits its applicability in industrial practice. Therefore, a vibration-based tool, which can reveal the gear wear and dynamic behaviours while requiring a small amount of experimental data for model parameter updating/calibration, is in vital need. Such a tool could bring significant benefits to gear wear monitoring and prediction in industrial practice.

2.5.2 Prediction of Surface Pitting Propagation

The research described in the previous sections focused on the prediction of the change in the gear teeth profile caused by abrasive wear. Some research [6, 70] involved predicting the propagation of fatigue pitting. In reference [6], an EHL model and a fatigue equation were combined to simulate the propagation behaviours of fatigue pitting and mild tooth profile change (caused by abrasive wear), under mixed EHL conditions. In this approach, the competitive behaviours between abrasive wear and fatigue pitting were successfully simulated during the gear wear propagation progression. However, as for this approach, high computational cost and high-level expert knowledge were required to realise the EHL model. Compared with the research described in reference [6], a more efficient approach was proposed by the researchers in reference [70], where the Archard wear model and the empirical fatigue pitting formula were used to predict both abrasive wear (in terms of wear depth) and fatigue pitting (in terms of surface pitting) propagation. Also, a statistical formulation was proposed in reference [86] to depict the evolution of asperity shape induced by wear and plastic deformations under mixed lubrication, and an asperity strain-hardening model was developed to predict the surface roughness change and the fatigue pitting propagation. Although the model predictions showed almost perfect agreement with the experimental reference measurements in reference [86], they must be further subjected to a larger number of experimental evaluations for a more decisive validation and a final judgment on their precision, which brings huge challenges to the application in industrial practices. Moreover, in references [6, 70, 86], none of their propagation processes was timely examined and calibrated by actual measurements to accommodate changes in operating and lubrication conditions, as well as wear conditions and rates.

In practice, the propagation rate of abrasive wear and fatigue pitting would be affected by contamination of the lubricant, change in surface roughness and changes in the operating conditions, etc. Therefore, without real-time examination and updating, the accuracy of predictions is uncertain and will decrease significantly as wear progresses. Therefore, reliable, effective, and efficient simulation methods are needed to predict the propagation of gear wear, with consideration of experimental/industrial measurements.

In addition to the above-mentioned physical model-based approaches, other kinds of techniques have also been developed and proven effective for fatigue pitting prediction. For example, the artificial neural network (ANN) was used in reference [87] to predict the severity of gear fatigue pitting. Based on some well-known standards, there are some methodologies developed for predicting fatigue pitting severity in references [88, 89]. The American Gear Manufacturers Association (AGMA) design standard was employed in reference [88] to predict pitting and bending fatigue crack initiation along the gear tooth profile. With the help of the ISO standard of gear micropitting (ISO/TR 15144-1:2020) and considering the operating load and speed conditions, a theoretical study was carried out in reference [89] to assess the risk of gear micropitting by determining the gear contact stress, sliding parameter, local contact temperature, and lubricant film thickness along the line of action of gear tooth contact. Note that a large set of experimental data is required to train the ANN or determine/optimise the parameters in the AGMA and ISO standards. In real applications and industry practice, it is hard to obtain sufficient historical data for training or parameter optimisation. Also, the ability to demonstrate the fatigue pitting propagation behaviours can help the analyst to better understand the fatigue mechanisms. However, ANN, AGMA, or ISO standards cannot reveal or exhibit the fatigue pitting propagation behaviours with details.

2.5.3 Research Gaps

From the above literature review on gear wear (including abrasive wear and fatigue pitting) prediction techniques, it can be seen that most of the existing wear prediction techniques are designed for predicting the tooth profile change induced by abrasive wear. In contrast, the researches on fatigue pitting prediction are rare. The possible reasons are as follows: (i) There are few effective and efficient models/tools for simulating fatigue pitting propagation behaviour; (ii) abrasive wear usually coexists during the fatigue pitting propagation progression, which brings in a complex surface degradation process and surface morphology; and (iii) the vibration feature induced by fatigue pitting is extremely complex, weak and difficult to extract. These challenges restrict the development of vibration-based methods for fatigue pitting propagation monitoring and prediction. Therefore, a reliable, effective, and efficient fatigue pitting model/tool is required to reveal and represent fatigue pitting propagation behaviours.

From the review of the existing research on gear wear prediction, it can also be found that the physics-based approach is widely used. Compared with the statistical model-based, artificial intelligence-based, and standard-based approaches, the physical (wear) model has its unique advantage, which is that the surface degradation behaviours can be represented to help understand the wear mechanism and its consequences for the gear system. However, the existing model-based wear prediction methodology has not been timely examined and calibrated using actual measurements, which could degrade the wear prediction accuracy. Therefore, it is necessary to develop a vibration-based tool for gear wear prediction, in which the wear model is developed with a complete understanding of the wear mechanism, and model parameters can be timely calibrated and updated with a reasonable amount of experimental data.

The digital twin (DT) is a virtual representation (mirror) of a physical structure or a system in real space along its lifecycle [90]. Through real-time interaction between the virtual model and physical structure, the degradation status of the system and its RUL can be reflected and evaluated effectively. Thanks to its unique speciality, DT has received considerable attention from the research community over the last decades. However, due to the complex structures and harsh operation conditions, research on DT-based gearbox transmission system RUL prediction is rather rare. Moreover, existing conceptual approaches [90–92] have limitations in indicating the specific contact statuses and providing insights on degradation stages of gearbox transmission systems, all of which are of high value to RUL prediction. Therefore, the development of a systematic and practical DT technology for gear wear monitoring and RUL prediction will benefit the research community and industrial practices significantly. It is the main research goal of this book.

2.6 SUMMARY

This chapter presents a review of vibration-based gear wear monitoring. From the literature review of vibration feature-based gear wear monitoring, it was found that most of the existing research focuses on tracking the abrasive wear-induced tooth profile change, which occurs at the millimetre level (macro-level wear). The researches for fatigue pitting monitoring (micro-level wear) and wear mechanism identification are rather rare and deserve more attention from the research community.

In reviewing the progress of vibration model-based gear wear monitoring, it can be seen that the Archard wear equation still plays an important role in modelling the abrasive wear behaviours. In contrast, models for fatigue pitting are limited, and the combination of the EHL model and fatigue criterion is the main approach to simulate fatigue pitting propagation behaviours, which is time-consuming and requires a high level of knowledge for the model establishment. Therefore, more effective and efficient models/tools for simulating fatigue pitting propagation behaviours are required.

Moreover, in practice, abrasive wear and fatigue pitting can both occur during the gear surface degradation process, appearing simultaneously or at different times on the same gear. Therefore, it is important that both the gear tooth profile change and surface pitting density can be monitored and predicted. To achieve this, there is a vital need to quantify the wear-induced tooth profile change (in terms of wear depth) and the surface pitting density in situations when these two wear events take place separately or simultaneously.

REFERENCES

[1] K. Feng, W.A. Smith, P. Borghesani, R.B. Randall, Z. Peng, Use of cyclostationary properties of vibration signals to identify gear wear mechanisms and track wear evolution, *Mechanical Systems and Signal Processing*, 150 (2021) 107258. https://doi.org/10.1016/j.ymssp.2020.107258.

[2] P. Rycerz, A. Kadiric, The influence of slide–roll ratio on the extent of micropitting damage in rolling–sliding contacts pertinent to gear applications, *Tribology Letters*, 67 (2019) 63. https://doi.org/10.1007/s11249-019-1174-7.

[3] H. Endo, *Simulation of gear faults and its application to the development of differential diagnostic technique*, School of Mechanical and Manufacturing Engineering, The University of New South Wales, Australia, 2005.

[4] American Gear Manufacturers Association, *Appearance of gear teeth: Terminology of wear and failure*, Vol. 40, American Gear Manufacturers Association, Virginia, United States, 1995.

[5] B. Bhushan, *Modern tribology handbook, two volume set*, CRC press, Boca Raton, United States, 2000.

[6] G.E. Morales-Espejel, P. Rycerz, A. Kadiric, Prediction of micropitting damage in gear teeth contacts considering the concurrent effects of surface fatigue and mild wear, *Wear*, 398–399 (2018) 99–115. https://doi.org/10.1016/j.wear.2017.11.016.

[7] H. Ding, A. Kahraman, Interactions between nonlinear spur gear dynamics and surface wear, *Journal of Sound and Vibration*, 307 (2007) 662–679. https://doi.org/10.1016/j.jsv.2007.06.030.

[8] R.B. Randall, A new method of modeling gear faults, *Journal of Mechanical Design*, 104 (1982) 259–267. https://doi.org/10.1115/1.3256334.

[9] I. Barányi, G. Kalácska, P. De Baets, Characterisation of abrasive worn surfaces by surface microtopography parameters, *International Journal of Sustainable Construction & Design*, 4(2) (2014). https://doi.org/10.21825/scad.v4i2.1036.

[10] F. Zhao, Z. Tian, X. Liang, M. Xie, An integrated prognostics method for failure time prediction of gears subject to the surface wear failure mode, *IEEE Transactions on Reliability*, 67 (2018) 316–327. https://doi.org/10.1109/TR.2017.2781147.

[11] W.D. Mark, *Performance-based gear metrology*, Wiley Online Library, Chichester, West Sussex, United Kingdom, 2012.

[12] R. Zhang, X. Gu, F. Gu, T. Wang, A.D. Ball, Gear wear process monitoring using a sideband estimator based on modulation signal bispectrum, *Applied Sciences*, 7 (2017) 274. https://doi.org/10.3390/app7030274.

[13] G.E. Totten, *ASM handbook, volume 18: Friction, lubrication, and wear technology*, ASM International, Cleveland, 1992.

[14] X. Zhang, W.A. Smith, P. Borghesani, Z. Peng, R.B. Randall, *Use of cyclostationarity to detect changes in gear surface roughness using vibration measurements, asset*

intelligence through integration and interoperability and contemporary vibration engineering technologies, Springer, Cham, 2019, pp. 763–771.

[15] Z.K. Zhu, Z.H. Feng, F.R. Kong, Cyclostationarity analysis for gearbox condition monitoring: Approaches and effectiveness, *Mechanical Systems and Signal Processing*, 19 (2005) 467–482. https://doi.org/10.1016/j.ymssp.2004.02.007.

[16] Y. Yang, W.A. Smith, P. Borghesani, Z. Peng, R.B. Randall, Detecting changes in gear surface roughness using vibration signals, in *Proceedings of the Acoustics Conference*, Hunter Valley, NSW, Australia, 2015, pp. 15–18.

[17] X. Zhang, J. Kang, J. Zhao, D. Cao, Features for fault diagnosis and prognosis of gearbox, *Chemical Engineering Transactions*, 33 (2013) 1027–1032.

[18] C. Hu, W.A. Smith, R.B. Randall, Z. Peng, Development of a gear vibration indicator and its application in gear wear monitoring, *Mechanical Systems and Signal Processing*, 76–77 (2016) 319–336. https://doi.org/10.1016/j.ymssp.2016.01.018.

[19] M. Amarnath, S.-K. Lee, Assessment of surface contact fatigue failure in a spur geared system based on the tribological and vibration parameter analysis, *Measurement*, 76 (2015) 32–44. https://doi.org/10.1016/j.measurement.2015.08.020.

[20] I. Yesilyurt, F. Gu, A.D. Ball, Gear tooth stiffness reduction measurement using modal analysis and its use in wear fault severity assessment of spur gears, *NDT & E International*, 36 (2003) 357–372. https://doi.org/10.1016/S0963-8695(03)00011-2.

[21] J. Mathew, J. Stecki, Comparison of vibration and direct reading ferrographic techniques in application to high-speed gears operating under steady and varying load conditions, *Lubrication Engineering*, 43 (1987) 646–653.

[22] S. Ziaran, R. Darula, Determination of the state of wear of high contact ratio gear sets by means of spectrum and cepstrum analysis, *Journal of Vibration and Acoustics*, 135 (2) (2013) 021008. https://doi.org/10.1115/1.4023208.

[23] C.J. Li, J.D. Limmer, Model-based condition index for tracking gear wear and fatigue damage, *Wear*, 241 (2000) 26–32. https://doi.org/10.1016/S0043-1648(00)00356-2.

[24] P. Kundu, A.K. Darpe, M.S. Kulkarni, A correlation coefficient based vibration indicator for detecting natural pitting progression in spur gears, *Mechanical Systems and Signal Processing*, 129 (2019) 741–763. https://doi.org/10.1016/j.ymssp.2019.04.058.

[25] F. Combet, L. Gelman, Novel adaptation of the demodulation technology for gear damage detection to variable amplitudes of mesh harmonics, *Mechanical Systems and Signal Processing*, 25 (2011) 839–845. https://doi.org/10.1016/j.ymssp.2010.07.008.

[26] A. Raad, J. Antoni, M. Sidahmed, Indicators of cyclostationarity: Theory and application to gear fault monitoring, *Mechanical Systems and Signal Processing*, 22 (2008) 574–587. https://doi.org/10.1016/j.ymssp.2007.09.011.

[27] H. Ozturk, I. Yesilyurt, M. Sabuncu, Detection and advancement monitoring of distributed pitting failure in gears, *Journal of Nondestructive Evaluation*, 29 (2010) 63–73. https://doi.org/10.1007/s10921-010-0066-4.

[28] A. Hase, H. Mishina, M. Wada, Correlation between features of acoustic emission signals and mechanical wear mechanisms, *Wear*, 292–293 (2012) 144–150. https://doi.org/10.1016/j.wear.2012.05.019.

[29] X. Chen, S. Wang, B. Qiao, Q. Chen, Basic research on machinery fault diagnostics: Past, present, and future trends, *Frontiers of Mechanical Engineering*, 13 (2018) 264–291. https://doi.org/10.1007/s11465-018-0472-3.

[30] N. Sawalhi, R.B. Randall, Simulating gear and bearing interactions in the presence of faults: Part I. The combined gear bearing dynamic model and the simulation of localised bearing faults, *Mechanical Systems and Signal Processing*, 22 (2008) 1924–1951. https://doi.org/10.1016/j.ymssp.2007.12.001.

[31] A. Al-Shyyab, A. Kahraman, A non-linear dynamic model for planetary gear sets, *Proceedings of the Institution of Mechanical Engineers, Part K: Journal of Multi-body Dynamics*, 221 (2007) 567–576. https://doi.org/10.1243/14644193JMBD92.

[32] M. Machado, P. Moreira, P. Flores, H.M. Lankarani, Compliant contact force models in multibody dynamics: Evolution of the Hertz contact theory, *Mechanism and Machine Theory*, 53 (2012) 99–121. https://doi.org/10.1016/j.mechmachtheory.2012.02.010.

[33] C. Zhou, H. Wang, An adhesive wear prediction method for double helical gears based on enhanced coordinate transformation and generalized sliding distance model, *Mechanism and Machine Theory*, 128 (2018) 58–83. https://doi.org/10.1016/j.mechmachtheory.2018.05.010.

[34] H. Wang, C. Zhou, Y. Lei, Z. Liu, An adhesive wear model for helical gears in line-contact mixed elastohydrodynamic lubrication, *Wear*, 426–427 (2019) 896–909. https://doi.org/10.1016/j.wear.2019.01.104.

[35] J. Wei, A. Zhang, P. Gao, A study of spur gear pitting under EHL conditions: Theoretical analysis and experiments, *Tribology International*, 94 (2016) 146–154. https://doi.org/10.1016/j.triboint.2015.08.037.

[36] A. Fernandez del Rincon, F. Viadero, M. Iglesias, P. García, A. de-Juan, R. Sancibrian, A model for the study of meshing stiffness in spur gear transmissions, *Mechanism and Machine Theory*, 61 (2013) 30–58. https://doi.org/10.1016/j.mechmachtheory.2012.10.008.

[37] K. Hayashi, Load distribution on the contact line of helical gear teeth: Part 1 fundamental concept, *Bulletin of JSME*, 6 (1963) 336–343.

[38] J. Boerner, Very efficient calculation of the load distribution on external gear sets – The method and applications of the program LVR, *Proc. of the 7th International Power Transmission and Gearing Conference*, ASME, San Diego, 1996, pp. 219–225.

[39] M. Pimsarn, K. Kazerounian, Efficient evaluation of spur gear tooth mesh load using pseudo-interference stiffness estimation method, *Mechanism and Machine Theory*, 37 (2002) 769–786. https://doi.org/10.1016/S0094-114X(02)00022-8.

[40] J.I. Pedrero, M. Pleguezuelos, M. Artés, J.A. Antona, Load distribution model along the line of contact for involute external gears, *Mechanism and Machine Theory*, 45 (2010) 780–794. https://doi.org/10.1016/j.mechmachtheory.2009.12.009.

[41] S. Li, Effect of addendum on contact strength, bending strength and basic performance parameters of a pair of spur gears, *Mechanism and Machine Theory*, 43 (2008) 1557–1584. https://doi.org/10.1016/j.mechmachtheory.2007.12.010.

[42] M. Arafa, M. Megahed, Evaluation of spur gear mesh compliance using the finite element method, *Proceedings of the Institution of Mechanical Engineers, Part C: Journal of Mechanical Engineering Science*, 213 (1999) 569–579. https://doi.org/10.1243/0954406991522509.

[43] Y. Temis, E. Kozharinov, D. Kalinin, Simulation of gear systems with dynamic analysis, in *Proceedings of the 14th IFToMM World Congress*, 2015, pp. 313–317.

[44] Z. Chen, Y. Shao, Mesh stiffness calculation of a spur gear pair with tooth profile modification and tooth root crack, *Mechanism and Machine Theory*, 62 (2013) 63–74. https://doi.org/10.1016/j.mechmachtheory.2012.10.012.

[45] J. Lin, C. Teng, E. Bergstedt, H. Li, Z. Shi, U. Olofsson, A quantitatively distributed wear-measurement method for spur gears during micro-pitting and pitting tests, *Tribology International*, 157 (2021) 106839. https://doi.org/10.1016/j.triboint.2020.106839.

[46] P. Bajpai, A. Kahraman, N.E. Anderson, A surface wear prediction methodology for parallel-axis gear pairs, *Journal of Tribology*, 126 (2004) 597–605. https://doi.org/10.1115/1.1691433.

[47] J. Brauer, S. Andersson, Simulation of wear in gears with flank interference – A mixed FE and analytical approach, *Wear*, 254 (2003) 1216–1232. https://doi.org/10.1016/S0043-1648(03)00338-7.

[48] A. Kahraman, H. Ding, A methodology to predict surface wear of planetary gears under dynamic conditions#, *Mechanics Based Design of Structures and Machines*, 38 (2010) 493–515. https://doi.org/10.1080/15397734.2010.501312.

[49] X. Liu, Y. Yang, J. Zhang, Investigation on coupling effects between surface wear and dynamics in a spur gear system, *Tribology International*, 101 (2016) 383–394. https://doi.org/10.1016/j.triboint.2016.05.006.

[50] J. Archard, Contact and rubbing of flat surfaces, *Journal of Applied Physics*, 24 (1953) 981–988.

[51] T. Osman, P. Velex, Static and dynamic simulations of mild abrasive wear in wide-faced solid spur and helical gears, *Mechanism and Machine Theory*, 45 (2010) 911–924. https://doi.org/10.1016/j.mechmachtheory.2010.01.003.

[52] M. Priest, C.M. Taylor, Automobile engine tribology – Approaching the surface, *Wear*, 241 (2000) 193–203. https://doi.org/10.1016/S0043-1648(00)00375-6.

[53] J.A. Brandão, P. Cerqueira, J.H.O. Seabra, M.J.D. Castro, Measurement of mean wear coefficient during gear tests under various operating conditions, *Tribology International*, 102 (2016) 61–69. https://doi.org/10.1016/j.triboint.2016.05.008.

[54] S. Wu, H.S. Cheng, Sliding wear calculation in spur gears, *Journal of Tribology*, 115 (1993) 493–500. https://doi.org/10.1115/1.2921665.

[55] M. Amarnath, C. Sujatha, S. Swarnamani, Experimental studies on the effects of reduction in gear tooth stiffness and lubricant film thickness in a spur geared system, *Tribology International*, 42 (2009) 340–352. https://doi.org/10.1016/j.triboint.2008.07.008.

[56] S. Li, *Lubrication and contact fatigue models for roller and gear contacts*, The Ohio State University, Columbus, Ohio, 2009.

[57] A. Flodin, S. Andersson, Simulation of mild wear in spur gears, *Wear*, 207 (1997) 16–23. https://doi.org/10.1016/S0043-1648(96)07467-4.

[58] A. Flodin, S. Andersson, Simulation of mild wear in helical gears, *Wear*, 241 (2000) 123–128. https://doi.org/10.1016/S0043-1648(00)00384-7.

[59] A. Flodin, S. Andersson, A simplified model for wear prediction in helical gears, *Wear*, 249 (2001) 285–292. https://doi.org/10.1016/S0043-1648(01)00556-7.

[60] A. Kahraman, P. Bajpai, N.E. Anderson, Influence of tooth profile deviations on helical gear wear, *Journal of Mechanical Design*, 127 (2004) 656–663. https://doi.org/10.1115/1.1899688.

[61] C. Yuksel, A. Kahraman, Dynamic tooth loads of planetary gear sets having tooth profile wear, *Mechanism and Machine Theory*, 39 (2004) 695–715. https://doi.org/10.1016/j.mechmachtheory.2004.03.001.

[62] S. Li, A. Kahraman, A transient mixed elastohydrodynamic lubrication model for spur gear pairs, *Journal of Tribology*, 132(1) (2010), 011501. https://doi.org/10.1115/1.4000270.

[63] S. Li, A. Kahraman, M. Klein, A fatigue model for spur gear contacts operating under mixed elastohydrodynamic lubrication conditions, *Journal of Mechanical Design*, 134(4) (2012), 041007. https://doi.org/10.1115/1.4005655.

[64] S. Li, A. Kahraman, A physics-based model to predict micro-pitting lives of lubricated point contacts, *International Journal of Fatigue*, 47 (2013) 205–215. https://doi.org/10.1016/j.ijfatigue.2012.09.002.

[65] K. Aslantaş, S. Taşgetiren, A study of spur gear pitting formation and life prediction, *Wear*, 257 (2004) 1167–1175. https://doi.org/10.1016/j.wear.2004.08.005.

[66] C.H. Chue, H.H. Chung, Pitting formation under rolling contact, *Theoretical and Applied Fracture Mechanics*, 34 (2000) 1–9. https://doi.org/10.1016/S0167-8442(00)00019-7.
[67] Y. Yin, Y.-X. Chen, L. Liu, Lifetime prediction for the subsurface crack propagation using three-dimensional dynamic FEA model, *Mechanical Systems and Signal Processing*, 87 (2017) 54–70. https://doi.org/10.1016/j.ymssp.2016.09.033.
[68] S. Li, A. Kahraman, A micro-pitting model for spur gear contacts, *International Journal of Fatigue*, 59 (2014) 224–233. https://doi.org/10.1016/j.ijfatigue.2013.08.015.
[69] G. Lundberg, Dynamic capacity of rolling bearings, *IVA Handlingar*, 196 (1947) 12.
[70] M. Weibring, L. Gondecki, P. Tenberge, Simulation of fatigue failure on tooth flanks in consideration of pitting initiation and growth, *Tribology International*, 131 (2019) 299–307. https://doi.org/10.1016/j.triboint.2018.10.029.
[71] S. Dhanasekaran, R. Gnanamoorthy, Gear tooth wear in sintered spur gears under dry running conditions, *Wear*, 265 (2008) 81–87. https://doi.org/10.1016/j.wear.2007.08.025.
[72] K.J. Sharif, H.P. Evans, R.W. Snidle, Prediction of the wear pattern in worm gears, *Wear*, 261 (2006) 666–673. https://doi.org/10.1016/j.wear.2006.01.018.
[73] M. Masjedi, M.M. Khonsari, On the prediction of steady-state wear rate in spur gears, *Wear*, 342–343 (2015) 234–243. https://doi.org/10.1016/j.wear.2015.08.010.
[74] M. Masjedi, M.M. Khonsari, Film thickness and asperity load formulas for line-contact elastohydrodynamic lubrication with provision for surface roughness, *Journal of Tribology*, 134(1) (2012), 011503. https://doi.org/10.1115/1.4005514.
[75] S. Li, Effects of machining errors, assembly errors and tooth modifications on loading capacity, load-sharing ratio and transmission error of a pair of spur gears, *Mechanism and Machine Theory*, 42 (2007) 698–726. https://doi.org/10.1016/j.mechmachtheory.2006.06.002.
[76] C. Xie, L. Hua, J. Lan, X. Han, X. Wan, X. Xiong, Improved analytical models for mesh stiffness and load sharing ratio of spur gears considering structure coupling effect, *Mechanical Systems and Signal Processing*, 111 (2018) 331–347. https://doi.org/10.1016/j.ymssp.2018.03.037.
[77] M.B. Sánchez, M. Pleguezuelos, J.I. Pedrero, Approximate equations for the meshing stiffness and the load sharing ratio of spur gears including hertzian effects, *Mechanism and Machine Theory*, 109 (2017) 231–249. https://doi.org/10.1016/j.mechmachtheory.2016.11.014.
[78] V. Hegadekatte, J. Hilgert, O. Kraft, N. Huber, Multi time scale simulations for wear prediction in micro-gears, *Wear*, 268 (2010) 316–324. https://doi.org/10.1016/j.wear.2009.08.017.
[79] H. Zhang, X. Shen, A dynamic tooth wear prediction model for reflecting 'two-sides' coupling relation between tooth wear accumulation and load sharing behavior in compound planetary gear set, *Proceedings of the Institution of Mechanical Engineers, Part C: Journal of Mechanical Engineering Science*, 234 (2020) 1746–1763. https://doi.org/10.1177/0954406219900085.
[80] G. Muthuveerappan, R. Thirumurugan, Prediction of theoretical wear in high contact ratio spur gear drive, *15th National Conference on Machines and Mechanisms*, Jharkkhand, India, 2011.
[81] D. Park, M. Kolivand, A. Kahraman, Prediction of surface wear of hypoid gears using a semi-analytical contact model, *Mechanism and Machine Theory*, 52 (2012) 180–194. https://doi.org/10.1016/j.mechmachtheory.2012.01.019.

[82] D. Park, M. Kolivand, A. Kahraman, An approximate method to predict surface wear of hypoid gears using surface interpolation, *Mechanism and Machine Theory*, 71 (2014) 64–78. https://doi.org/10.1016/j.mechmachtheory.2013.09.002.

[83] S. Sun, S. Wang, Y. Wang, T.C. Lim, Y. Yang, Prediction and optimization of hobbing gear geometric deviations, *Mechanism and Machine Theory*, 120 (2018) 288–301. https://doi.org/10.1016/j.mechmachtheory.2017.09.002.

[84] S. Raadnui, K. Wangwira, Abrasive Spur Gear Wear Prediction Model: Utilization of the Statistical Design Experiments, in *The 30th International Congress on Condition Monitoring and Diagnostic Engineering Management*, Preston, United Kingdom, 2018.

[85] X. Zhao, X. Xu, W. Zhao, Z. Jiang, X. Liu, Trend prediction of wear fault of wind generator high-speed gear using a fusion of UICA and PE method, *Cluster Computing*, 20 (2017) 427–437. https://doi.org/10.1007/s10586-017-0733-7.

[86] R. Guilbault, S. Lalonde, A stochastic prediction of roughness evolution in dynamic contact modelling applied to gear mild wear and contact fatigue, *Tribology International*, 140 (2019) 105854. https://doi.org/10.1016/j.triboint.2019.105854.

[87] A. Krishnakumari, A. Perumal, R.C. Panda, Prediction for condition monitoring of wear and film thickness in a gear box, Engineering Mechanics, 19(6) (2012) 393–405.

[88] P. Kumar, H. Hirani, A. Agrawal, Fatigue failure prediction in spur gear pair using AGMA approach, *Materials Today: Proceedings*, 4 (2017) 2470–2477. https://doi.org/10.1016/j.matpr.2017.02.099.

[89] I.S. Al-Tubi, H. Long, Prediction of wind turbine gear micropitting under variable load and speed conditions using ISO/TR 15144-1: 2010, *Proceedings of the Institution of Mechanical Engineers, Part C: Journal of Mechanical Engineering Science*, 227 (2013) 1898–1914. https://doi.org/10.1177/0954406212469593.

[90] H. Jiang, S. Qin, J. Fu, J. Zhang, G. Ding, How to model and implement connections between physical and virtual models for digital twin application, *Journal of Manufacturing Systems*, 58 (2021) 36–51. https://doi.org/10.1016/j.jmsy.2020.05.012.

[91] Q. Qi, F. Tao, T. Hu, N. Anwer, A. Liu, Y. Wei, L. Wang, A.Y.C. Nee, Enabling technologies and tools for digital twin, *Journal of Manufacturing Systems*, 58 (2021) 3–21. https://doi.org/10.1016/j.jmsy.2019.10.001.

[92] J. Ma, H. Chen, Y. Zhang, H. Guo, Y. Ren, R. Mo, L. Liu, A digital twin-driven production management system for production workshop, *The International Journal of Advanced Manufacturing Technology*, 110 (2020) 1385–1397. https://doi.org/10.1007/s00170-020-05977-5.

3 Methodology

3.1 THE OVERALL STRATEGY OF THE VIBRATION-BASED INTEGRATED SYSTEM FOR GEAR WEAR MONITORING

The integrated system for vibration-based gear wear monitoring and prediction comprises two distinct stages:

Stage 1: Identification of the wear mechanism utilising advanced vibration-based techniques.
Stage 2: Monitoring and prediction of wear propagation through vibration-based approaches.

The project is structured around four specific research objectives: Objective 1 falls under Stage 1, while Objectives 2 through 4 are part of Stage 2.

Objective 1: Identification of gear wear mechanism and tracking wear evolution through the analysis of cyclostationary properties of measured vibrations.
Objective 2: Dynamic model development.
Objective 3: Monitoring and prediction of tooth profile changes during wear progression.
Objective 4: Development of a digital twin approach for monitoring and prediction of surface pitting and tooth profile changes.

Objective 3 only focuses on monitoring and predicting one wear phenomenon: tooth profile change (e.g. caused by abrasive wear). Objective 4 enhances the work of Objective 3 by monitoring and predicting two wear phenomena/events, namely tooth profile change and surface pitting. In practical applications, it is common that multiple wear events co-exist during gear wear progression. Therefore, Objective 4 accounts for these two common wear phenomena: The propagation of surface pitting and changes in tooth profile. To clarify the connections between the objectives mentioned above, a schematic diagram illustrating the overall strategy for gear wear monitoring and prediction is presented in Figure 3.1. The specific techniques utilised to realise each objective will be introduced in Sections 3.3 and 3.4. Also,

DOI: 10.1201/9781003686736-3

the congruent relationship between each objective and the following chapters of this book is shown in Figure 3.1.

To conduct the research stated in Stage 1, and also to demonstrate and validate the effectiveness of the developed vibration-based integrated system in gear wear

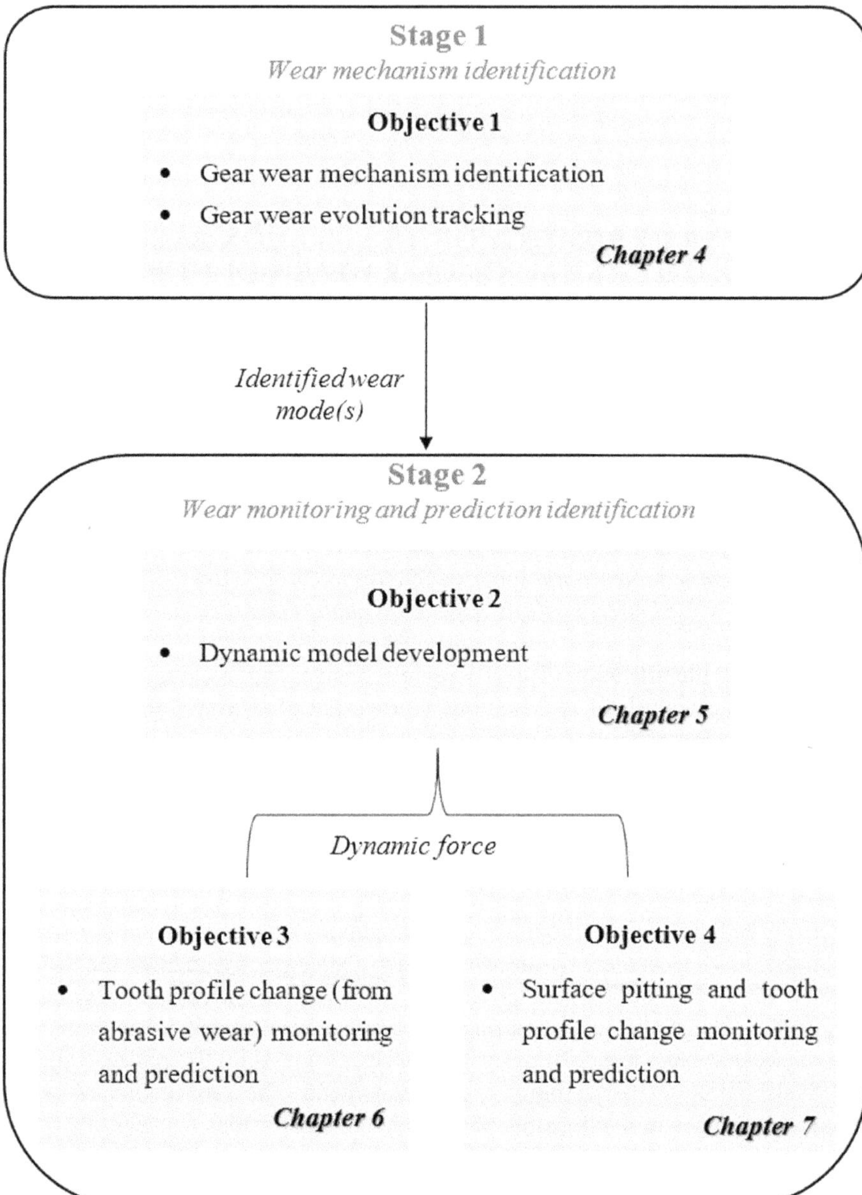

FIGURE 3.1 The schematic diagram of the book methodology. Made by the author.

monitoring and prediction, two endurance tests were performed on a spur gearbox at UNSW, conducted under different lubrication conditions to generate different dominant wear phenomena/events. The experimental research facilities and test programmes designed to achieve the objectives outlined in this book are provided in the following section.

3.2 EXPERIMENTAL RESEARCH FACILITIES AND TEST PROGRAMMES

3.2.1 Spur Gearbox at University of New South Wales

A single-stage spur gearbox rig, illustrated in Figure 3.2, was utilised to perform gear wear tests. The gearbox consists of an input shaft and an output shaft, each carrying two modular gears with a module size of 2, having 19 and 52 teeth, respectively. To accelerate wear rate, the gears were made of mild steel (JIS S45C) which was not

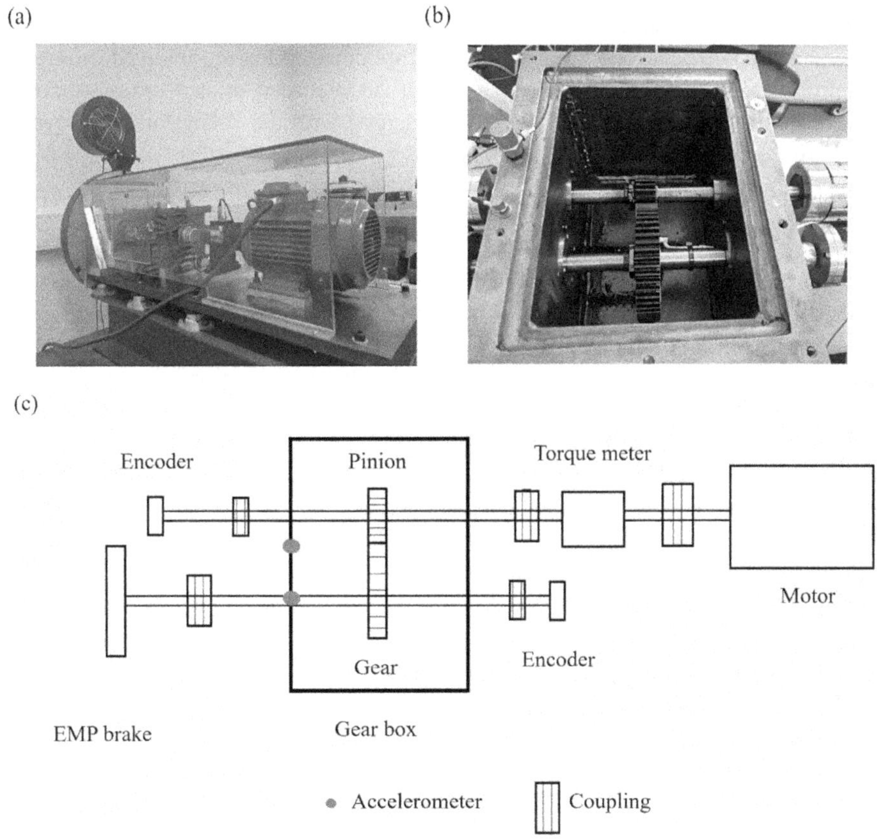

FIGURE 3.2 The spur gear test rig at the University of New South Wales (UNSW) [1]. Made by the author.

subjected to heat treatment, resulting in a hardness of less than 194 HB. The precision grade of the gears is JIS grade 4 (JIS B1702:1976), with standard full depth. The input shaft is driven by a 4-kW electric motor, whose instantaneous rotational speed is controlled by a variable frequency drive (VFD). The motor is connected to a torque meter that can monitor the instantaneous torque of the gear transmission system. An electromagnetic particle (EMP) brake is connected to the output shaft at its end and serves to control the torque transmitted by the gear transmission system. Two encoders are installed at the remaining free ends of the shafts. The connections of the gearbox shafts with the motor, brake, and encoders are facilitated by couplings that have high torsional stiffness and low bending stiffness.

Two vibration sensors (B&K 4396 and B&K 4394 accelerometers) are mounted on the top of the gearbox casing at the positions shown in Figure 3.2(b). The sensitivity of the B&K 4396 accelerometer is $10.0\,\mathrm{mV}/(\mathrm{ms}^{-2})$ with a nominal frequency range of 1–14000 Hz, whereas the sensitivity and frequency range of the B&K 4394 are $1.00\,\mathrm{mV}/(\mathrm{ms}^{-2})$ and 1–25000 Hz, respectively. The lubrication is provided by an oil bath, and the kinematic viscosity of the oil is 146 (40 °C, Shell Spirax S2 A 80 W-90). Before the lubricated test and the unlubricated test, three teeth on each gear (six teeth in total) were randomly selected and marked to monitor the evolution of wear on the gear teeth surfaces. The moulding procedure consisted of applying Microset 101 thixotropic silicone polymer to a series of marked gear teeth at each stoppage during the endurance tests. The moulds were then tagged with the collection time (cycles) and stored for further analysis.

3.2.2 Wear Tests and Data Collection

Two run-to-failure tests were conducted to simulate gear wear progression behaviours under different lubrication conditions. A test with lubrication was conducted to generate fatigue pitting on the gear surfaces, accompanied by mild tooth profile changes due to the abrasive wear. Subsequently, a test without lubrication was conducted to generate abrasive wear, resulting in severe tooth profile changes with little to no surface pitting observed. Further details of these two tests are provided below.

The lubricated test was performed to simulate the natural progression of fatigue pitting propagation. It was conducted using a pre-roughened gear pair to accelerate the degradation process, with abrasive wear occurring initially when the gear tooth surface was rough. The initial gear surface condition/morphology is shown in Figure 3.3. The lubricated test ran for a total of 3.25 million cycles of the pinion (the driving gear). The test rig was stopped approximately every 0.1 million cycles to record the surface condition of the six gears by using a moulding technique, as shown in Figure 3.4. During this endurance test, the motor torque was set to 20 Nm for the entire duration, which is approximately nine times the pinion's rated torque for surface durability, ensuring the occurrence of surface pitting. During short specific intervals within the test campaign (mostly before each stoppage), the input shaft rotational speed was adjusted to 2 Hz, 6 Hz, 10 Hz, 16 Hz, and 20 Hz from the pre-set operating speed to capture vibration and tacho/encoder signals. Each speed adjustment lasted around 10 seconds, after which the input shaft speed was returned to the pre-set operating speed. This speed variation was intended to generate different types of transmission error signals, with the relevant investigations reported in reference [2]. Vibration signals

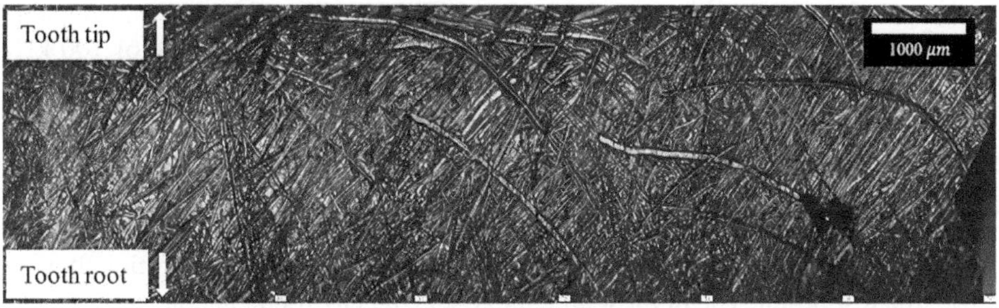

FIGURE 3.3 Initial surface of sandpaper-roughened pinion gear [1]. Photograph by the author.

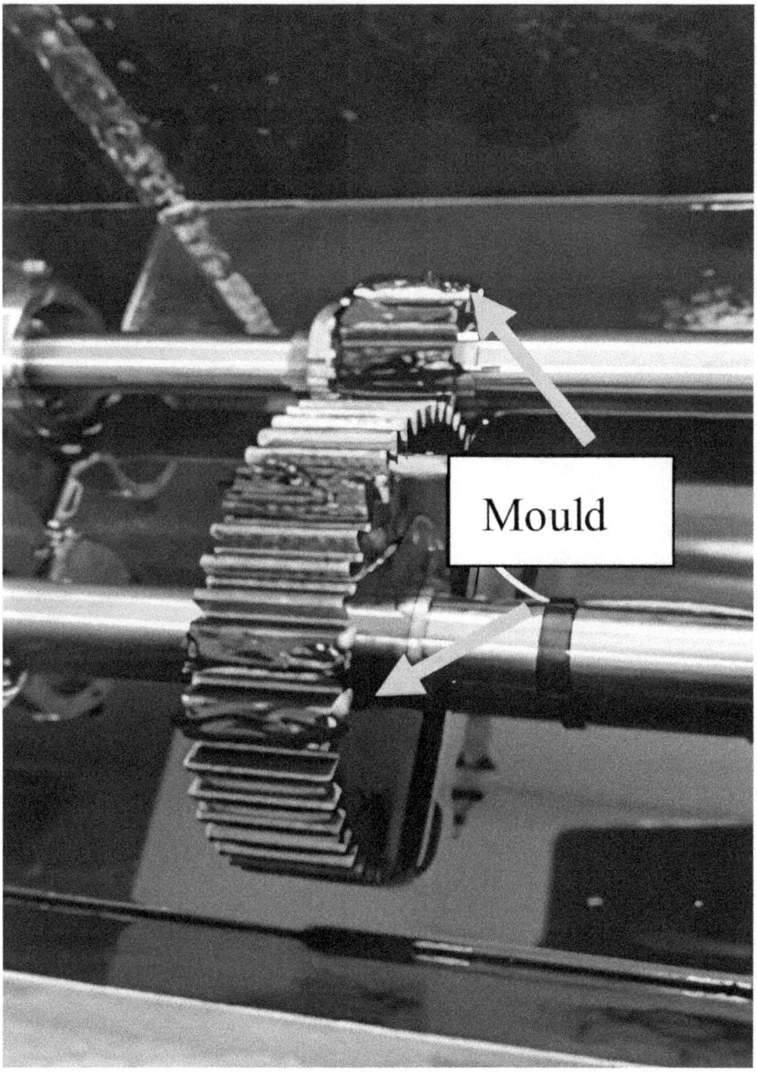

FIGURE 3.4 Gear pair and mould making (in lubricated test) [1]. Photograph by the author.

were regularly acquired (approximately every 6000 cycles) within the test campaign, with a record length of 10 seconds and a corresponding sampling rate of 100 kHz. Figure 3.5 shows an image of the pinion and an image of a pinion tooth surface after the test with lubrication, illustrating the occurrence of fatigue pitting.

The test without lubrication (dry test) was conducted with a new pair of gears over approximately 38,000 cycles of the pinion. The test rig was stopped approximately every 5,000 cycles to record the surface condition using the moulding technique. Figure 3.6 shows the surface of a tooth on the pinion, which was not artificially worn before the test began. The dry test aimed to create an environment conducive to high rates of abrasive wear, resulting in rapid tooth profile changes, but without significant levels of pitting. To achieve this, the motor torque was set to 5 Nm, and the input shaft speed to 10 Hz for the entire duration of the test. Similarly, during short specific intervals within the test campaign (mostly before each stoppage), the input shaft rotational speed was adjusted to 2 Hz from the pre-set operating speed to capture vibration and tacho/encoder signals. The vibration signals were sampled at 100 kHz for a duration of 10 seconds. Wear particles were collected during the dry test using adhesive paper, as shown in Figure 3.7, and their mass was used to calculate the average wear depth. Figure 3.8 shows the appearance of the pinion and its gear surface after the dry test.

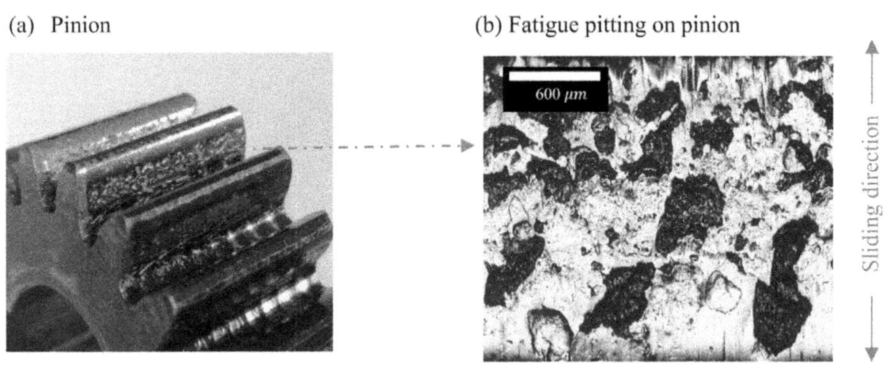

FIGURE 3.5 Gear surfaces after the lubricated test [1]. Photograph by the author.

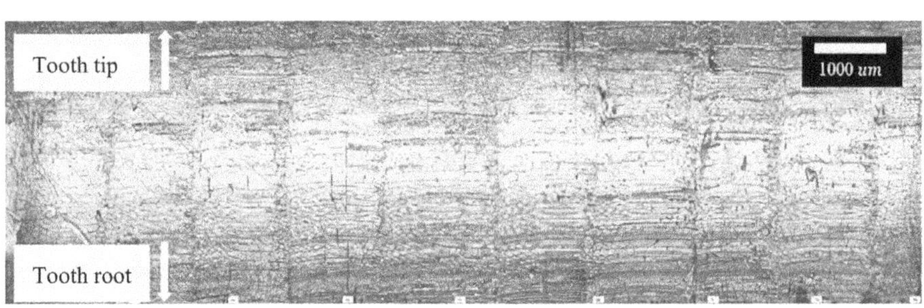

FIGURE 3.6 Initial pinion surface before the dry test. Photograph by the author.

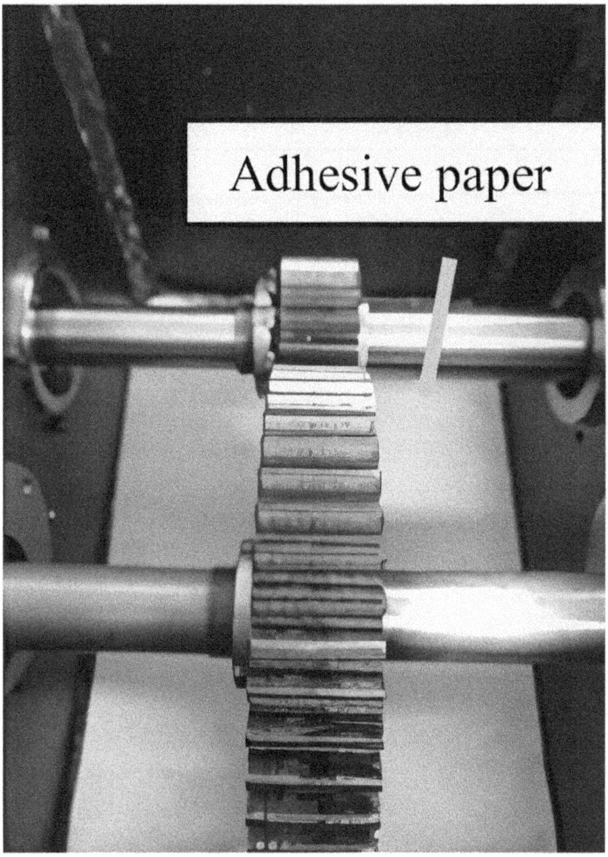

FIGURE 3.7 Gear pair and adhesive paper in dry test [3]. Photograph by the author.

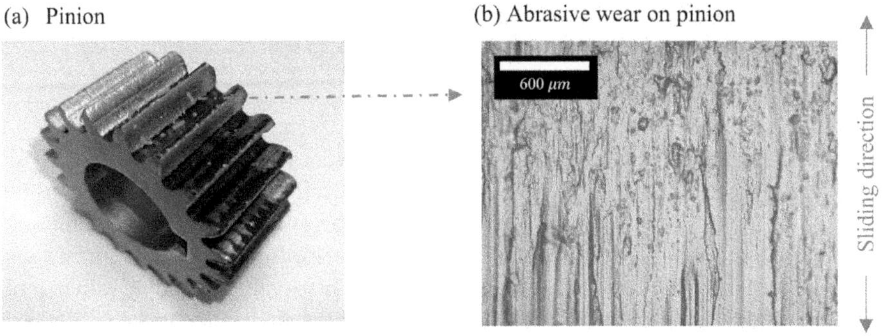

FIGURE 3.8 Gear surfaces after the dry test [1]. Photograph by the author.

TABLE 3.1
Settings of Lubricated Test and Dry Test

Indicators	Test 1	Test 2
Lubricated	Yes	No
Pre-worn	Yes	No
Max cycles	3,250,000 cycles	38,000 cycles
Stop intervals	100,000 cycles	5,000 cycles
Load	20 Nm	5 Nm
Input shaft speed (constant)	10 Hz	10 Hz

It should be emphasised that the aim of the two tests is not to conduct the most realistic wear progression possible, nor to investigate the effects of speed and load on wear rates. The objective is rather to generate significant abrasive wear in one test and substantial fatigue pitting in the other, so that techniques for monitoring and predicting the progression of these wear types can be developed and tested.

The lubricated test and dry test are employed in this book to help achieve the above-mentioned objectives 1, 3, and 4. More specifically, two different dominant wear mechanisms/events occurred in the lubricated test and dry test, respectively: Fatigue pitting and abrasive wear. The measured vibrations in the two tests can be utilised to develop the vibration-based techniques for wear mechanism identification, which corresponds to objective 1 of this book. Additionally, as for objectives 2, 3, and 4, the measured vibrations are used for model calibration, followed by comparisons with simulations to update model parameters for predicting gear wear progression. These aspects will be introduced in Section 3.3 and Section 3.4. The details and differences between the lubricated test and dry test are summarised in Table 3.1.

3.3 GEAR WEAR MECHANISM IDENTIFICATION AND WEAR EVOLUTION TRACKING USING CYCLOSTATIONARY PROPERTIES OF VIBRATIONS (OBJECTIVE 1)

It is essential to identify the gear wear mechanism and understand its impacts on gear teeth before monitoring and predicting the wear propagation, so that the corresponding maintenance strategies for specific wear events can be developed and scheduled. According to the literature review in Section 2.3, it can be observed that the existing approaches for wear mechanism identification are off-line techniques. For instance, a widely used approach is the visual inspection of a worn surface and/or the wear particles generated from the surface. This method requires either interruption of the operation (in the case of visual inspection) or a delay in analysing wear debris, and the operating status of gearboxes cannot be promptly reflected. Therefore, it is crucial to identify wear mechanisms (abrasive wear and fatigue pitting) using an efficient and effective online approach and track their wear evolution.

In general, changes in the profile of gear teeth lead to changes in the deterministic component of vibrations. In contrast, the sliding-induced random vibrations from the asperity contacts between two mating gears are closely related to the surface morphology [4, 5]. From the perspective of surface features, abrasive wear tends to alter the gear tooth profile, while fatigue pitting has a negligible effect on the gear tooth profile unless the fatigue pitting is extremely severe [6]. However, both abrasive wear and fatigue pitting change the micro-geometry of gear tooth surfaces. Therefore, the sliding-induced random vibrations contain information related to both abrasive wear and fatigue pitting.

However, in the measured vibrations, the sliding-induced random vibrations are usually mixed with the signals from other unrelated vibration-generating mechanisms or white noise. This creates difficulties in extracting surface morphology-related information for wear mechanism identification and wear severity tracking, and also reduces the effectiveness of traditional indicators (e.g. RMS and kurtosis) in gear wear monitoring. To effectively extract the sliding vibrations from background noise (i.e. the signals from other unrelated vibration-generating mechanisms or white noise), the special characteristics of sliding vibrations were investigated in references [7, 8]. It was found that the sliding vibration has second-order cyclostationarity (CS2) due to the time-varying sliding velocity and contact force on the tooth pairs. Therefore, the cyclostationary properties of vibrations and relevant techniques, such as spectral coherence maps and an indicator of CS2, are employed in this book to identify the wear mechanism and then track its propagation.

The main procedures to achieve objective 1 in this book are as follows. Firstly, the relationship between the spatial frequency of the surface morphology and the frequency of sliding vibrations is investigated and established. Then, fatigue pitting and abrasive wear are identified and separated using the carrier frequency of measured vibrations and ICS2 of vibrations [9] (an indicator of CS2). Fatigue pitting information is carried in the low-frequency range, while abrasive wear information is carried in the high-frequency range. Finally, the propagation of fatigue pitting and abrasive wear is monitored and tracked using ICS2 within the appropriate frequency bands. A schematic diagram of the vibration-based wear mechanism identification and evolution tracking is shown in Figure 3.9. Further details of this development can be found in Chapter 4.

3.4 MODEL-BASED GEAR WEAR MONITORING AND PREDICTION METHODOLOGY

Once the wear mechanism is identified, gear wear propagation can be monitored and predicted using the corresponding approach/technique, allowing the remaining useful life (RUL) of the gear system to be estimated. As reviewed in Section 2.4, compared with statistical models and artificial intelligence, the physical model-based wear monitoring methodology offers several advantages, such as requiring fewer measurements for training or calibration purposes and providing an in-depth understanding of the wear mechanism. Therefore, model-based techniques are utilised in this book for monitoring and predicting gear wear propagation.

However, in practice, during gear wear progression, the wear propagation rate tends to vary due to factors such as oil contamination, changes in operating conditions, and changes in surface roughness. Therefore, to ensure accurate prediction results, regular comparisons between simulations and actual measurements are scheduled in

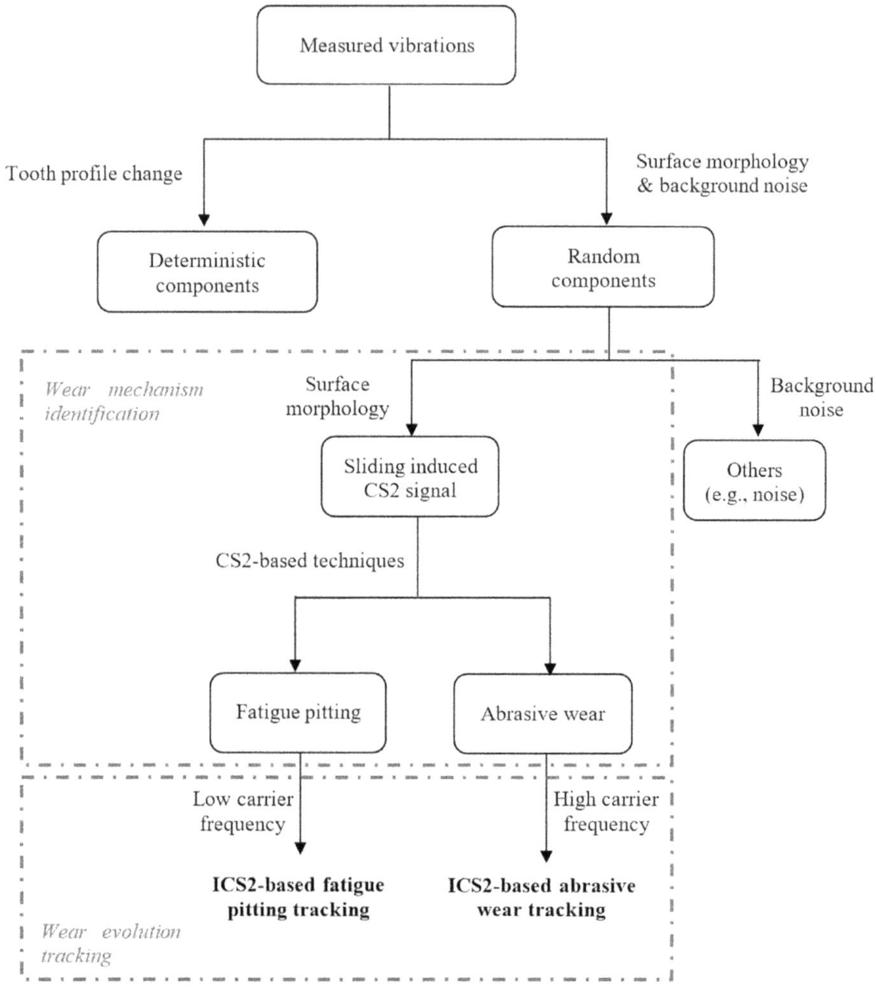

FIGURE 3.9 Vibration-based wear identification and evolution tracking [1]. Made by the author.

this project to update the model parameters if necessary. This updating procedure addresses the wear rate shift issue and guarantees that the best possible prediction of RUL can be achieved at any particular time. The framework of the model-based wear prediction proposed in this project is shown in Figure 3.10. In the following sections, specific models and techniques for monitoring different wear phenomena will be introduced. It should be emphasised that the main novelty of this model-based gear wear monitoring scheme is that regular comparisons between simulations and actual measurements are conducted to update the model parameters whenever

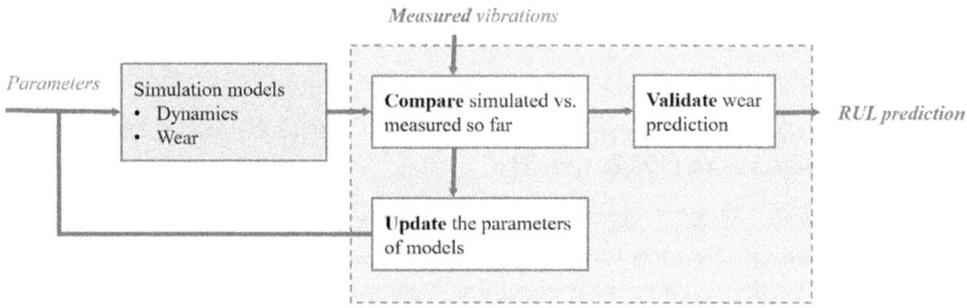

FIGURE 3.10 Model-based gear wear monitoring and prediction. Made by the author.

necessary, ensuring that an accurate prediction of wear propagation can be achieved. Additionally, compared with wear particle measurements or images obtained through scanning the worn tooth profile, vibration offers a unique advantage, as it can be easily and quickly obtained for updating purposes without interrupting the operation of the gearbox.

3.4.1 Spur Gearbox Dynamic Model Development (Objective 2)

From the literature review of Section 2.4.1, modelling the contact force is a critical procedure for gear wear monitoring and prediction, as it is an important input of tribological (wear) models. Most of the existing research uses finite element models and empirical equations to calculate the contact force between the meshing gears, with the obtained contact force being under quasi-static conditions. However, in practice, the gear system typically operates under dynamic conditions, and due to inertial effects, dynamic meshing forces are typically larger than the corresponding quasi-static forces, with significantly different waveforms [10]. Therefore, using quasi-static force can lead to noticeable errors in subsequent wear prediction. Compared with simple finite element models and empirical equations, the dynamic model provides a contact force under dynamic conditions that accounts for the inertia effects [11–13]. Therefore, the dynamic model is chosen to provide the necessary inputs to wear models and generate simulated vibrations in this book.

To achieve objective 2 of this book, a 21-degree-of-freedoms (DOFs) lumped parameter dynamic model is established based on the UNSW gearbox test bench (as shown in Figure 3.2). The model includes the motor, shafts, gears, casing, and couplings. The basic motion equations of the dynamic model are derived based on Newton's Second Law of Motion.

To ensure the outputs (such as dynamic contact forces and vibrations) from the dynamic model are sufficiently close to the measurements from the experimental rig, the model has been validated and calibrated through a series of tests, including impact tests, speed ramp tests, and several constant speed tests. Further details of the dynamic model development can be found in Chapter 5. Note that the 21-DOFs lumped parameter dynamic model is developed based on the fixed-axis spur gearbox test rig shown in Figure 3.2, which has a different structure and properties from the spur gearbox test rig used in previously published work [11–13].

3.4.2 Monitoring and Prediction of Tooth Profile Changes from Abrasive Wear (Objective 3)

With the dynamic contact force obtained from the established dynamic model, the wear progression can be monitored and predicted using specific tribological (wear) models. Tooth profile change is a common wear phenomenon, and it is typically caused by abrasive wear [6]. The tooth profile alteration can cause stress concentration, significantly increasing the risk of tooth breakage. Therefore, it is essential to monitor and predict the tooth profile changes due to abrasive wear.

The physical-model-based approach is chosen in this book to monitor the tooth profile changes due to abrasive wear. Therefore, a tribological (wear) model which can represent the abrasive wear propagation behaviours and their induced tooth profile change is required. As reviewed in Section 2.4.2, although many advanced wear models have been proposed using various methodologies and parameter sets, the Archard wear model [14] remains the most commonly used for practical applications, and is chosen in this book. However, none of the previously published research [10, 15–18] considered the contribution of adjacent contact points to wear depth accumulation in the Archard wear model, which can affect wear distribution and impair the wear prediction accuracy. Therefore, in this book, the Archard wear model is improved to calculate the wear depth on the gear tooth, considering the effect of Hertzian deformation, which provides a contact area rather than a line, and including the effects from adjacent contact points.

To solve the previously discussed issue of changing wear propagation rates, simulation and measured vibrations are compared regularly to update the model parameters if necessary, and a vibration-based scheme for updating gear wear prediction is proposed in this book. The overall procedure of the proposed methodology is shown in Figure 3.11. The modelling component of this methodology consists of two interacting simulation models: A dynamic model and an Archard wear model. Based on the input of the gear tooth profile geometry, the dynamic model predicts gear tooth dynamic contact forces, which are passed on to the Archard wear model to estimate the gear wear distribution. This, in turn, modifies the gear tooth profile geometry, which is then fed back into the dynamic model. This iterative loop enables a knowledge-based prediction of gear wear propagation. To address the issue of changing wear rates (due to factors such as oil contamination, changes in operating conditions, changes in surface roughness, etc.), vibrations from the gearbox dynamic model are compared with measured vibrations to track the quality of the wear model predictions and, if necessary, update the gear wear model parameters. This updating procedure is shown on the right of Figure 3.11. As shown in the results presented in Chapter 6, the vibration-based updating scheme can deliver reliable and accurate gear tooth profile change monitoring and prediction results from abrasive wear. Further details of the development can be found in Chapter 6.

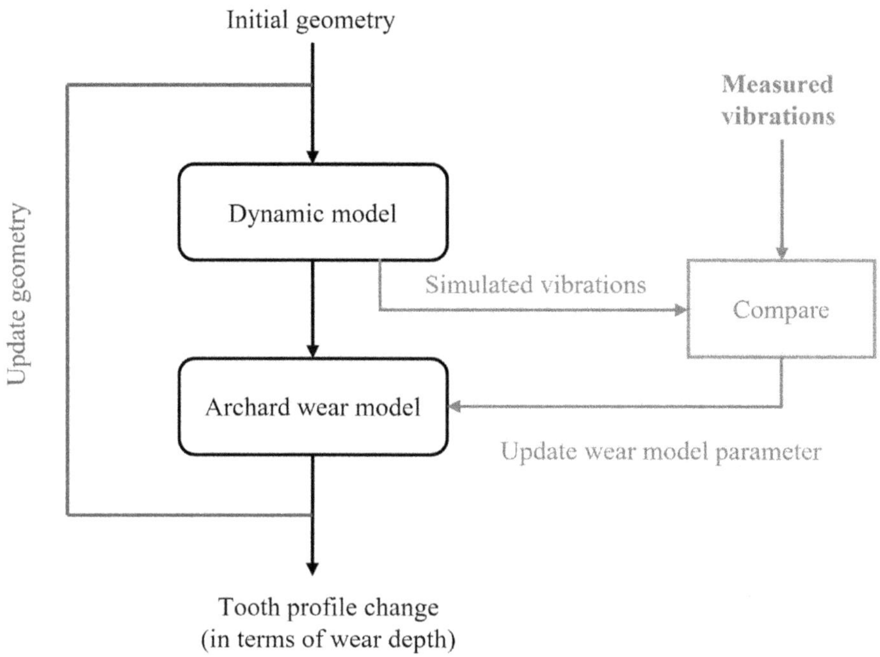

FIGURE 3.11 Procedures for vibration-based abrasive wear prediction [19]. Made by the author.

3.4.3 Monitoring and Prediction of Surface Pitting Propagation and Tooth Profile Change from Abrasive Wear Using a Digital Twin Approach (Objective 4)

Fatigue pitting is another common wear phenomenon during the gear service life. It is induced by fatigue under cyclic loading, which can result in large valleys on the gear tooth surface. However, the effective working tooth profile (considered across the entire face width) often remains unchanged, unless pitting is extremely severe [6]. Consequently, fatigue pitting differs fundamentally from abrasive wear. In addition, fatigue pitting could promote the generation of surface spalls across the full tooth width, reducing surface durability and potentially leading to tooth breakage [20]. Therefore, it is essential to monitor and predict the surface pitting propagation, which supports accurate RUL estimation of the gear system.

Based on the literature review in Section 2.5.2, most of the existing approaches for modelling surface pitting propagation combine an elastohydrodynamic lubrication (EHL) model with a fatigue criterion [21]. These methods require substantial computational resources and detailed knowledge of the surface micro-geometry and lubrication conditions due to the incorporation of surface roughness, prohibiting their widespread application in practice. To address this issue, this book develops a simple and efficient surface pitting model to simulate and predict the pitting propagation behaviours based on the Lundberg–Palmgren model [22].

However, similar to the previously discussed issue of changes in the abrasive wear rate, the surface pitting propagation rate would also be affected by a variety of factors, such as lubrication quality and quantity, contact pressure distribution, surface roughness, and operating conditions, all of which may change significantly. Without frequent monitoring and, if necessary, updating of the wear model parameters, the accuracy of the prediction results cannot be guaranteed and is likely to decrease significantly during the surface pitting propagation.

To accurately monitor and predict the surface pitting propagation, a similar vibration-based updating scheme as introduced in Section 3.4.2 is employed in this book, as shown in Figure 3.12. In the proposed vibration-based updating scheme, contact pressure from the dynamic model is provided to the developed fatigue pitting model, and the surface pitting propagation can be predicted in terms of pitting density. The occurrence and progression of surface pitting could reduce the contact area of the mating gear teeth and subsequently alter its contact pressure, which could impact further surface pitting propagation. This knowledge-based surface pitting propagation is illustrated on the left side of Figure 3.12. To guarantee accurate predictions, regular updating of the pitting coefficients is implemented by comparing with measured vibrations (ICS2), when available, as on the right side of Figure 3.12. More details of the development of fatigue and surface pitting propagation prediction can be found in Chapter 7.

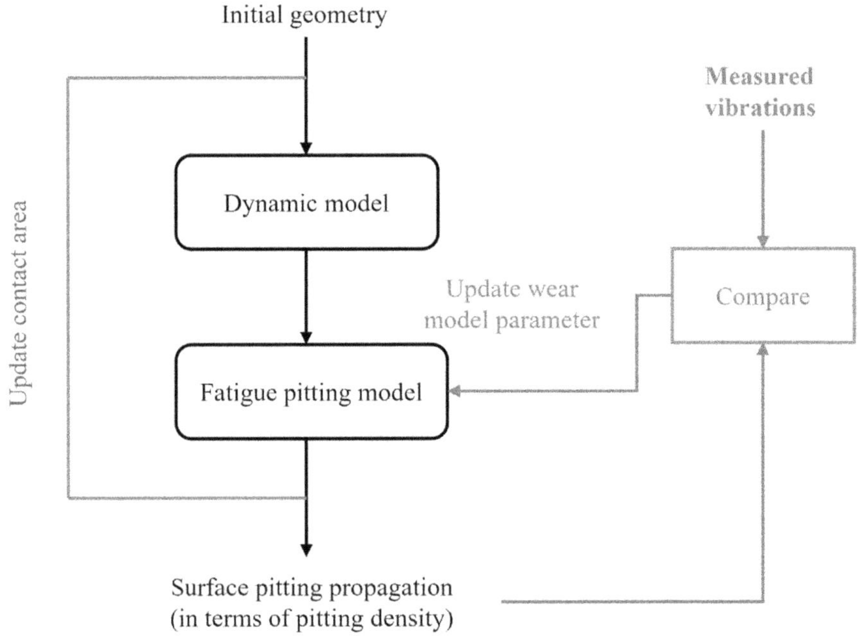

FIGURE 3.12 Procedures of the vibration-based fatigue pitting propagation predictions. Made by the author.

In practice, during the propagation of fatigue pitting, abrasive wear may co-exist due to oil contamination. As a result, both abrasive wear-induced tooth profile changes and surface pitting can occur simultaneously or at different times as part of the surface degradation process in the same gear. These processes can significantly influence one another. Therefore, it is essential to develop a reliable and efficient tool, which can utilise the capability of efficient physics-based models and measurements to predict the surface pitting propagation and the tooth profile change caused by the co-existing abrasive wear.

To address the aforementioned issue, in this book, a digital twin approach for monitoring and prediction of surface pitting and tooth profile changes from abrasive wear is developed, whose skeleton is shown in Figure 3.13. Based on the input of gear tooth profile geometry and operating conditions, the dynamic model can provide dynamic contact forces and simulate vibrations of the gear system. The contact pressure is then calculated using the Hertzian contact theory. With the contact pressure as an input, the Archard wear model can estimate the abrasive wear rate and consequently predict the gear tooth profile geometry at a specified future time, which is subsequently fed back into the dynamic model. Meanwhile, the contact pressure is passed on to a surface pitting model to predict the surface pitting density, which consequently modifies the contact area and then alters the Hertzian contact pressure. To obtain accurate predictions, the wear coefficients are regularly updated by comparing measured vibrations, when available, with the simulations. Further details on the development of the digital twin approach for monitoring and predicting both surface pitting propagation and tooth profile changes can be found in Chapter 7.

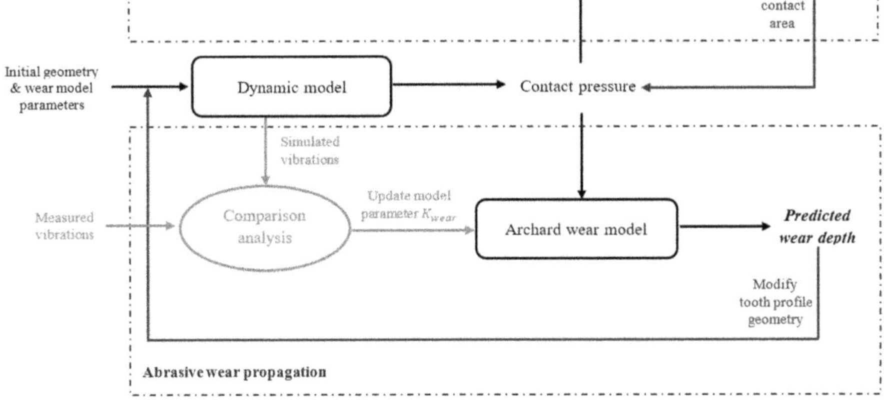

FIGURE 3.13 Digital twin approach for monitoring and prediction of surface pitting and tooth profile changes from abrasive wear [3]. Made by the author.

REFERENCES

[1] K. Feng, W.A. Smith, P. Borghesani, R.B. Randall, Z. Peng, Use of cyclostationary properties of vibration signals to identify gear wear mechanisms and track wear evolution, *Mechanical Systems and Signal Processing*, 150 (2021) 107258. https://doi.org/10.1016/j.ymssp.2020.107258.

[2] Z.Y. Chin, W.A. Smith, P. Borghesani, R.B. Randall, Z. Peng, Absolute transmission error: A simple new tool for assessing gear wear, *Mechanical Systems and Signal Processing*, 146 (2021) 107070. https://doi.org/10.1016/j.ymssp.2020.107070.

[3] K. Feng, W.A. Smith, R.B. Randall, H Wu, Z. Peng, Vibration-based monitoring and prediction of surface profile change and pitting density in a spur gear wear process, *Mechanical Systems and Signal Processing*, 165 (2022) 108319. https://doi.org/10.1016/j.ymssp.2021.108319.

[4] A. Soom, J.-W. Chen, Simulation of random surface roughness-induced contact vibrations at hertzian contacts during steady sliding, *Journal of Tribology*, 108 (1986) 123–127. https://doi.org/10.1115/1.3261131.

[5] S. Li , A. Kahraman, A tribo-dynamic model of a spur gear pair, *Journal of Sound and Vibration*, 332 (2013) 4963–4978. https://doi.org/10.1016/j.jsv.2013.04.022.

[6] American Gear Manufacturers Association, *Appearance of gear teeth: Terminology of wear and failure*, Vol 40, *American Gear Manufacturers Association*, Virginia, United States, 1995.

[7] X. Zhang, W.A. Smith, P. Borghesani, Z. Peng, R.B. Randall, *Use of cyclostationarity to detect changes in gear surface roughness using vibration measurements, asset intelligence through integration and interoperability and contemporary vibration engineering technologies*, Springer, Cham, Switzerland, 2019, pp. 763–771.

[8] Y. Yang, W.A. Smith, P. Borghesani, Z. Peng, R.B. Randall, *Detecting changes in gear surface roughness using vibration signals*, in Proceedings of the Acoustics Conference, Hunter Valley, NSW, Australia, 2015, pp. 15–18.

[9] A. Raad, J. Antoni, M. Sidahmed, Indicators of cyclostationarity: Theory and application to gear fault monitoring, *Mechanical Systems and Signal Processing*, 22 (2008) 574–587. https://doi.org/10.1016/j.ymssp.2007.09.011.

[10] H. Ding, A. Kahraman, Interactions between nonlinear spur gear dynamics and surface wear, *Journal of Sound and Vibration*, 307 (2007) 662–679. https://doi.org/10.1016/j.jsv.2007.06.030.

[11] H. Endo, *Simulation of gear faults and its application to the development of differential diagnostic technique*, School of Mechanical and Manufacturing Engineering, The University of New South Wales, Australia, 2005.

[12] X. Chen, S. Wang, B. Qiao, Q. Chen, Basic research on machinery fault diagnostics: Past, present, and future trends, *Frontiers of Mechanical Engineering*, 13 (2018) 264–291. https://doi.org/10.1007/s11465-018-0472-3.

[13] L. Deshpande, N. Sawalhi, R. Randall, *Gearbox fault simulation using finite element model reduction technique*, in Proceedings of Acoustics, Gold Coast, Australia, 2011, p. 111.

[14] J. Archard, Contact and rubbing of flat surfaces, *Journal of Applied Physics*, 24 (1953) 981–988.

[15] P. Bajpai, A. Kahraman, N.E. Anderson, A surface wear prediction methodology for parallel-axis gear pairs, *Journal of Tribology*, 126 (2004) 597–605. https://doi.org/10.1115/1.1691433.

[16] T. Osman, P. Velex, Static and dynamic simulations of mild abrasive wear in wide-faced solid spur and helical gears, *Mechanism and Machine Theory*, 45 (2010) 911–924. https://doi.org/10.1016/j.mechmachtheory.2010.01.003.

[17] A. Flodin, S. Andersson, Simulation of mild wear in spur gears, *Wear*, 207 (1997) 16–23. https://doi.org/10.1016/S0043-1648(96)07467-4.
[18] Z. Shen, B. Qiao, L. Yang, W. Luo, X. Chen, Evaluating the influence of tooth surface wear on TVMS of planetary gear set, *Mechanism and Machine Theory*, 136 (2019) 206–223. https://doi.org/10.1016/j.mechmachtheory.2019.03.014.
[19] K. Feng, P. Borghesani, W.A. Smith, R.B. Randall, Z.Y. Chin, J. Ren, Z. Peng, Vibration-based updating of wear prediction for spur gears, *Wear*, 426–427 (2019) 1410–1415. https://doi.org/10.1016/j.wear.2019.01.017.
[20] P.J.L. Fernandes, C. McDuling, Surface contact fatigue failures in gears, *Engineering Failure Analysis*, 4 (1997) 99–107. https://doi.org/10.1016/S1350-6307(97)00006-X.
[21] G.E. Morales-Espejel, P. Rycerz, A. Kadiric, Prediction of micropitting damage in gear teeth contacts considering the concurrent effects of surface fatigue and mild wear, *Wear*, 398–399 (2018) 99–115. https://doi.org/10.1016/j.wear.2017.11.016.
[22] G. Lundberg, Dynamic capacity of rolling bearings, *IVA Handlingar*, 196 (1947) 12.

4 Gear Wear Identification and Evolution Tracking Using Vibration Cyclostationarity

4.1 INTRODUCTION

Fatigue pitting is a fatigue-induced material loss, after which the effective working tooth profile (considered across the entire face width) often remains unchanged (unless pitting is extremely severe). In general, fatigue pitting originates from a subsurface crack and occurs initially at the dedendum of the gear tooth or near the pitch line, due to the repetitive rolling-sliding contact and high contact stress [1, 2]. In contrast, abrasive wear, usually caused by particle contamination or insufficient lubrication [3], is the removal of material (often across the entire face width) induced by sliding contact, and each piece of material removed contributes to a change in the profile [4]. Fatigue pitting and abrasive wear have different effects on the gear tooth surface and profile geometry. Therefore, it is essential to identify and monitor fatigue pitting and abrasive wear separately for wear prediction, so that appropriate maintenance decisions can be made to avoid catastrophic failures, unexpected economic losses, and serious accidents.

Vibration analysis is an effective approach to monitor the health state of machines [5–8], and it has been well established for the detection and diagnosis of common gear faults, such as gear tooth root cracks [9–11], surface spalls [12–15], and tooth breakage [16–19]. In contrast, vibration-based techniques for wear analysis, including wear mechanism identification and wear tracking, are rather rare. One reason for this is that there is a coupling effect between gear wear and gear dynamic characteristics, which produces complex gear dynamic characteristics/vibration features [20, 21], making it difficult to extract wear-related vibration features and develop specific vibration-based indicator(s) for wear identification and monitoring.

From the published research work, it can be found that most of the existing vibration-based gear wear monitoring techniques were designed for detecting and tracking gear tooth profile change (accumulated gear wear), which is usually caused by abrasive wear. More specifically, the existing vibration-based techniques use gear meshing harmonics or signal 'power,' which is mainly based on the deterministic components of vibration, to monitor the gear wear process in terms of tooth profile change. Since fatigue pitting does not change the tooth profile noticeably (unless pitting is extremely severe) [4], it has a negligible effect on the deterministic components of vibration, which are closely related to gear macro-geometry, i.e.

DOI: 10.1201/9781003686736-4

the tooth profile. Therefore, these proposed techniques for processing deterministic components of vibration have a very limited ability to identify gear wear mechanisms and track fatigue pitting severity.

In contrast, the random signal components induced by varying sliding motions and contact forces contain useful information for gear wear mechanism identification and wear severity tracking because they are closely related to gear surface morphology characteristics [22, 23]. Furthermore, even though gear wear (fatigue pitting and abrasive wear) is usually likely to be uniform around the gear, it is still random from tooth to tooth. Therefore, studying the random signal components offers significant potential for advancing gear wear mechanism identification and the development of wear severity monitoring techniques. However, there remains limited research on the random signal components modulated by gear mesh behaviours and on how these components might be used for wear monitoring.

In recently published research [24, 25], the relationship between gear surface roughness and sliding-induced random vibration was investigated. In the gear transmission system, all gear operations involve some form of sliding contact between the mating gear surfaces, and this sliding contact can produce random vibrations from the asperity contacts between the gears, the nature of which is dependent on a number of factors, such as the speed and load, the lubrication conditions, and the microgeometry of the surfaces. Therefore, the sliding-induced vibrations have random characteristics. With this as a basis, second-order cyclostationarity was found in the sliding-induced vibrations due to the time-varying sliding velocity [26] and contact force on the gear tooth pairs in reference [25]. A positive relationship between gear tooth surface roughness level and the degree of second-order cyclostationarity (CS2) of vibration was found. However, the connection was found by a later investigation [24] to be more complex, based on a wider range of roughness values and a longer experimental duration. To date, insufficient conclusions could be drawn in evaluating the cyclostationary features for vibration-based gear wear mechanism identification and wear severity tracking.

Although the cyclostationarity of vibration signals has been widely studied, most published works have focused on signal modulation characteristics, using the 'degree' of cyclostationarity (e.g. a measure of the level of amplitude modulation) to correlate with fault severity [27–29], while studies on the spectral frequency of cyclostationary vibration at gear mesh cyclic frequency are rare. In other words, properties associated with the cyclic frequency α have been widely studied, while those of the carrier (spectral) frequency f have not, despite both having clear physical underpinnings. In reference [24], the authors used the mean carrier frequency to investigate the impacts of speed on the vibration signals modulated at the gear mesh frequency. Although a direct ratio between the mean carrier frequency and running speed was found, no further investigation was undertaken; for instance, the relationship between the carrier frequency and gear surface morphology was not investigated.

Inspired by the achievements reported in references [24, 25], this chapter aims to present a vibration-based approach for gear wear mechanism identification, and then to track wear evolution with the help of a cyclostationary indicator, the combination of which was not investigated previously. This work presented in this chapter is based on the observation that sliding vibration signals are significantly influenced by surface features induced by different wear mechanisms and also affected by modulation effects caused by sliding velocity.

4.2 HYPOTHESIS AND PROPOSED VIBRATION-BASED APPROACH FOR GEAR WEAR IDENTIFICATION

In this section, the surface feature differences arising from fatigue pitting and abrasive wear are introduced and summarised, followed by a brief description of micro-level gear surface feature effects on sliding-induced vibrations. A hypothesis for gear wear mechanism identification will then be presented, after which a vibration-based wear mechanism identification approach is presented.

4.2.1 Surface Feature Differences and Their Effects on Sliding Vibrations

As mentioned before, from a macro-scale point of view, abrasive wear tends to introduce a change to the gear tooth profile, while fatigue pitting has negligible effects on the gear tooth profile unless the fatigue pitting is extremely severe [4]. Both abrasive wear and fatigue pitting change the micro-geometry of gear tooth surfaces, that is, the surface morphologies on a micrometre scale, as demonstrated in Figure 4.1 (abrasive wear) and Figure 4.2 (fatigue pitting). The images were obtained using a moulding procedure outlined in reference [30]. Abrasive wear causes scratches or gouges on the tooth surface that are oriented in the direction of sliding. As a result, the worn, rough surfaces have micro-scaled morphological features in short wavelengths and high spatial frequencies [31], while profile

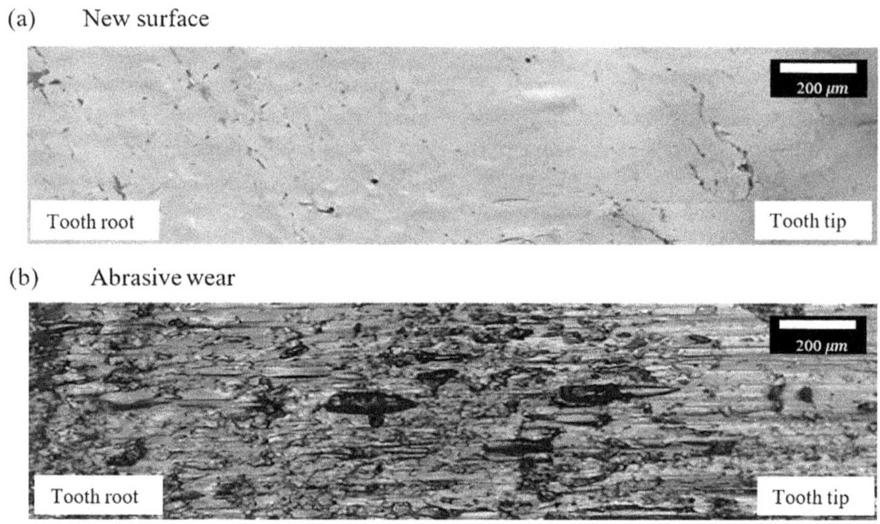

FIGURE 4.1 Micro-scale changes in gear surface morphology during the abrasive wear process. Photograph by the author [33].

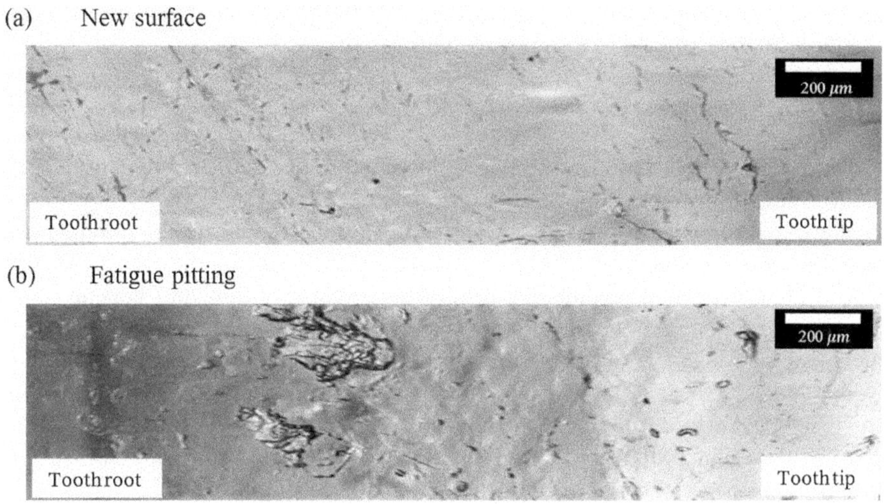

FIGURE 4.2 Micro-scale changes in gear surface morphology during the fatigue pitting process. Photograph by the author [33].

changes are long wavelengths. Fatigue pitting will instead introduce large valleys on the gear tooth surface, which contain longer wavelengths and lower spatial frequencies [32] in comparison with surface roughness changes associated with the abrasive wear process. These surface feature differences can help in separating fatigue pitting and abrasive wear.

For gear systems, the surfaces of contacting gear teeth are subject to combined rolling and sliding action as the gears rotate. Since rolling resistance is considerably smaller than the sliding resistance [34], usually, its contribution to the total tooth friction is ignored, and only the sliding-induced friction is considered. During the gear meshing process, the sliding-induced tooth friction will result in the generation of vibration, which is a random signal and is closely related to the gear surface morphology [23]. Although the sliding-induced vibration might have very low energy compared with the vibration induced by macro-geometric effects (such as tooth profile change) of gears, it contains rich information about the surface morphology. It has been proposed that the surface morphology information can be detected and extracted through cyclostationary tools, as demonstrated in reference [25]. Therefore, the use of cyclostationary properties of sliding-induced vibration should be able to identify different wear mechanisms and then track their evolution. This possibility will be further explored and investigated in the following sections.

4.2.2 Hypothesis for Wear Mechanism Identification

The sliding-induced vibration from the asperity contacts between two mating gears is closely related to the surface morphology. Different surface morphologies will result in different frequency components of sliding-induced vibration. The relationship between the spatial frequency of the surface morphology and the frequency of sliding vibration can be assumed to follow the relationship:

$$f_v \propto v_s \cdot f_s \tag{4.1}$$

where f_v (Hz) is the dominant sliding induced vibration frequency, v_s (m/s) is the sliding velocity of the mating gear surface, and f_s (1/m) is the dominant spatial frequency of the roughness of the mating surfaces. Despite the presence of multiple phenomena (e.g. the dependency of power on load, transfer function effects), it is at least expected that, even if not quantitatively following this relationship, the frequency range characterising vibration due to sliding will grow with sliding velocity and spatial frequency of the roughness of the mating surfaces. The sliding velocity v_s is determined by the gear kinematics and remains virtually unchanged under mild/medium wear (micro-level). Surfaces modified in an abrasive wear process or fatigue pitting process have different spatial frequencies, as shown in Figure 4.3 and Figure 4.4, respectively.

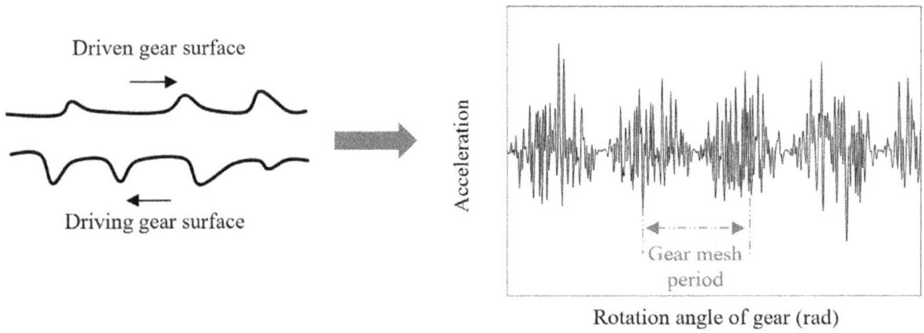

FIGURE 4.3 The fatigue pitting induced sliding vibration characteristics. Made by the author [33].

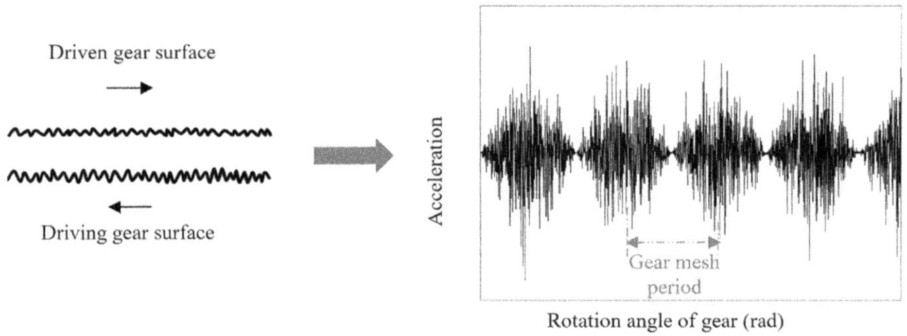

FIGURE 4.4 The abrasive wear-induced sliding vibration characteristics. Made by the author [33].

This is expected to affect the spectral content (carrier frequency band) of the resulting gear mesh-cyclic CS2 components. Based on the above assumptions, we expect that:

1) sliding of fatigue pitted surfaces, exhibiting low spatial frequency, should generate gear mesh-cyclic CS2 components with low spectral frequencies (Figure 4.3).
2) sliding of worn surfaces subject to abrasion, characterised by surface morphologies with high spatial frequency, should result in gear mesh-cyclic CS2 components with high spectral frequencies (Figure 4.4).

It should be noted that in the above discussion, Eq. (4.1) , Figure 4.3 and Figure 4.4 are used to describe the manner of excitation, and that the measured response would be shaped by a system transfer function, and so the observed dominant carrier frequencies in the CS content of the measured signal would also depend on system resonances, as well as the cyclostationary tool used for their detection. While the theoretical excitation frequency (band) would be directly proportional to speed and surface spatial frequency, the observed vibration carrier frequency (band) is likely to appear to move discretely between dominant resonances. The fact that system resonances are generally not of uniform strength further complicates this phenomenon. To mitigate this issue, 'normalised' cyclostationary tools such as cyclic-coherence [35] can be chosen to investigate the spectral content of the signals.

Based on this physical intuition, analysing the relative carrier frequency range of sliding-induced vibrations is a promising approach for identifying fatigue pitting and abrasive wear. This hypothesis will be validated using experimental data in Section 4.3.

4.2.3 Vibration-Based Approach for Wear Mechanism Identification

To summarise the previous discussion, it is expected that (i) the CS2 vibration components with gear mesh cyclic frequency are symptomatic of changes in tooth surface morphology, and (ii) the dominant spectral (carrier) frequency f of those CS2 components can be used to determine the key wavelengths of surface alterations and therefore the dominant wear mechanism.

With these hypotheses, a vibration-based approach for gear wear mechanism identification is illustrated in Figure 4.5. In the vibration signal, only the random components with second-order cyclostationarity (CS2), with a cyclic frequency corresponding to a particular gear mesh, are considered for gear wear (fatigue pitting and abrasive wear) identification. The deterministic ones are expected to be instead correlated to macroscopic profile changes (more severe wear effects). The proposed vibration-based wear identification approach will be validated using experimental data in the following section.

It must be noted that the scope of this work has been restricted to micro-surface alteration and therefore to the random components of vibration, but future developments could then combine this information with that extracted from the deterministic component for a full picture of the tooth degradation.

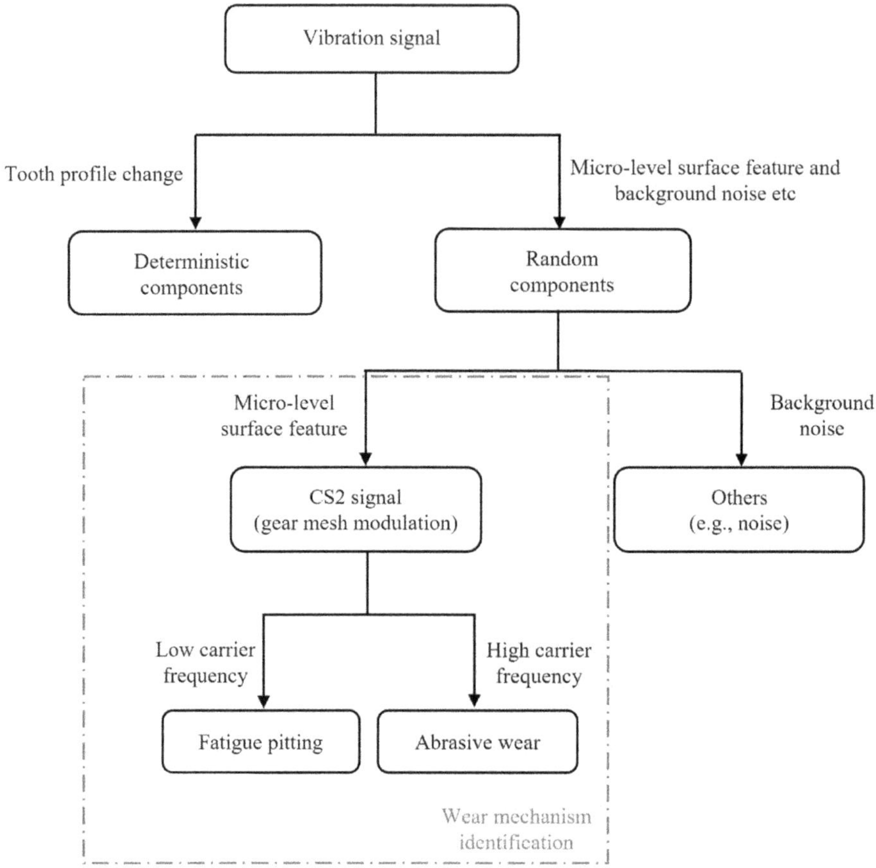

FIGURE 4.5 Diagram of vibration-based gear wear mechanism identification approach [33]. Made by the author.

4.3 OBSERVATIONS IN GEAR SYSTEMS

4.3.1 Tribological Features Used to Describe Fatigue Pitting and Abrasive Wear Propagation

It was observed through visual inspection that wear (fatigue pitting and abrasive wear) occurred quite uniformly on the gears and that the pinion was worn much more severely than the driven (big) gear due to the gear ratio. An optical microscope with a 5× magnification objective lens was used to capture 2D images of the gear moulds to monitor the wear progress qualitatively. A laser scanning confocal microscope (LSCM) was used to capture 3D images of the gear moulds for quantitative analysis. The captured 3D images contain the height and spatial information of the gear tooth surfaces for numerical characterisations of changes in the surface morphologies during the wear processes.

For fatigue pitting, initially, both the number and size of the pits increase, until the pitting becomes quite severe, at which point nearby pits tend to join, and the actual number of pits may decrease. Thus, the most robust indicator to describe pitting severity is probably the pitted area, and in Section 4.4, this is used to assess the performance of the proposed technique in monitoring pitting propagation.

For abrasive wear, a high wear rate often results in a surface deviation from a perfect involute (macroscopic profile change). Meanwhile, at the micro-level, the gear tooth surfaces tend to become rougher when abrasive wear propagates (very short wavelength) [36]. Surface roughness S_a, the arithmetic average of absolute values of surface deviation from the mean surface level, is often used to characterise the surface change at the micro-metre level. In this book, surface roughness S_a is applied as the tribological reference to help check the capability of the vibration-based indicators/techniques in monitoring the abrasive wear-induced micro-surface feature change, which will be introduced in Section 4.4.

4.3.2 Theory of Vibration-Based Wear Mechanism Identification Techniques

In this section, the hypothesis and proposed approach in Sections 4.2.2 and 4.2.3 for wear mechanism identification are introduced in detail.

To obtain the sliding induced vibrations, the deterministic/random signal separation technique, time synchronous averaging (TSA) or discrete/random separation (DRS) [37–41], is applied to remove the deterministic components of the vibrations. The remainder is referred to as the residual signal. Spectral coherence, defined in reference [35], is employed to qualitatively explore the spectral frequency distribution of sliding-induced vibration at gear mesh cyclic frequency, and to compute what will be referred to as 'mean carrier frequency' (i.e. the first moment of the spectral coherence along the spectral frequency axis). The definition of spectral coherence is:

$$\gamma_x(\alpha, f) = \frac{S_x(\alpha, f)}{\sqrt{S_x(0, f) S_x(0, f-\alpha)}} \tag{4.2}$$

where $S_x(\alpha, f)$ represents the ordinary power spectral density at frequency f, i.e. the CS content at frequency f is normalised by the power at frequencies f and $f-\alpha$in the stationary part of the signal.

Although the deterministic components of vibrations are removed using TSA, in the residual signal, the sliding-induced vibration is still mixed with the background noise, which brings difficulties in extracting surface morphology-related information for wear mechanism identification and wear severity tracking.

CS2 signals have a close relationship with surface morphology [24, 25, 42, 43]. The general hypothesis is that when the surface roughness increases due to the occurrence of fatigue pitting or abrasive wear, the friction between the mating gear surfaces will increase, resulting in a stronger sliding-induced vibration, while the gear mesh modulation pattern remains unchanged. Therefore, it is hypothesised that the second-order cyclostationarity of the sliding-induced vibration will increase correspondingly, as qualitatively depicted in Figure 4.6.

ICS2 [44, 45], an indicator to measure the degree of second-order cyclostationarity, can be used to assess the phenomenon illustrated in Figure 4.6, relative to stationary background noise. The definition of ICS2 is:

$$\mathrm{ICS2}^{\mathcal{A}_h,\mathrm{H}} = \frac{\sum_{h=1:H} \max_{n \in \mathcal{A}_h} \left(\mathrm{SES}[n]^2 \right)}{\mathrm{SES}[0]^2} \tag{4.3}$$

where $\mathcal{A}_h$ with h=1 is a set of cyclic frequencies of interest (with a tolerance band in the case of expected cyclic frequency deviations), and $\mathcal{A}_h$ ($h = 2,\ldots,$ H) represents the equivalent sets for the corresponding harmonics. H indicates the maximum gear mesh harmonics to be taken into consideration. In the case of gears, to monitor gear wear progression, $\mathcal{A}_1$ is set as the gear mesh frequency with three times the cyclic resolution as the tolerance band. SES refers to the squared envelope (amplitude) spectrum [6]. Note that the background noise is a stationary signal because any signal value (event) is equally probable to happen given any other signal value (another event) at any two time instances, no matter how far apart they are.

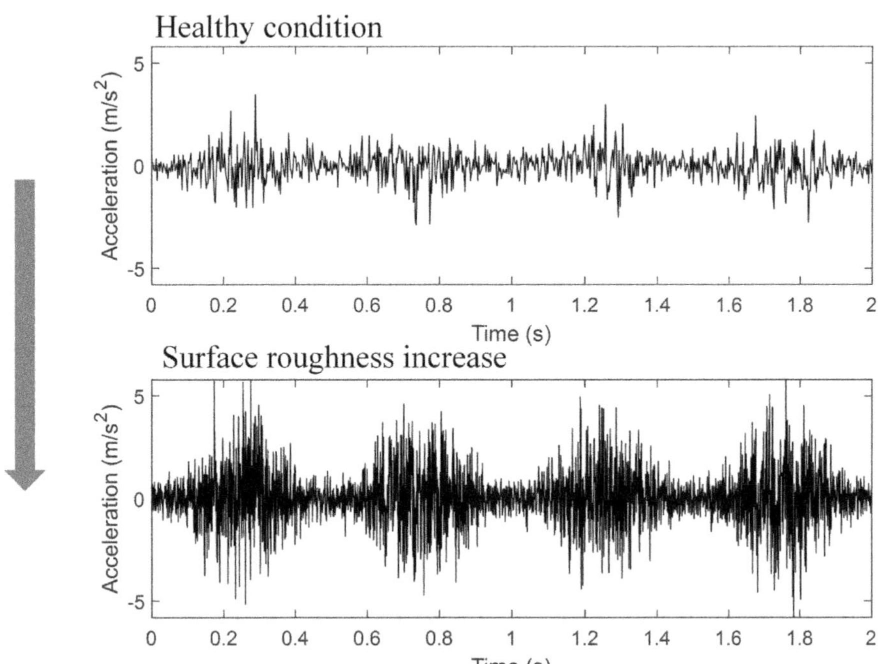

FIGURE 4.6 Diagram to illustrate the hypothesis: Increase in second-order cyclostationarity due to an increase in surface roughness. Made by the author [33].

By using a 1/3-binary tree filter bank (or other similar decomposition representation) [46], ICS2 can be used to select one band with maximum cyclostationarity, in which the influences of background noise become less significant, and the vibration characteristics related to surface morphology are enhanced [45].

Based on the hypothesis in Section 4.2.2, the ICS2-based frequency band selection results will be different for fatigue pitting and abrasive wear, and can be used to separate them. Further, because background noise is minimised in the selected frequency band, the ICS2 level within the selected band should have better performance in tracking the evolution of fatigue pitting/abrasive wear severity compared with the full-band ICS2 level. The effectiveness of ICS2-based wear mechanism identification and wear monitoring is validated in Section 4.3.3 and Section 4.4, respectively, using experimental data measured from gear systems.

4.3.3 Observation Results

In this section, the observation results in gear systems using the above-mentioned techniques, spectral coherence map and ICS2-based band selection, are presented to validate the hypothesis for wear mechanism identification proposed in Section 4.2 and Section 4.3.2.

4.3.3.1 Observation Results of the Lubricated Test

Figure 4.7 shows the distribution in the spectral coherence of the gear mesh-cyclic CS2 component (cyclic frequency) for the lubricated test at 10 Hz, in which fatigue pitting is the dominant wear mechanism. The spectral coherence software developed in reference [47] was used for this analysis. From Figure 4.7, it can be found that the carrier frequency distribution is quite different before and after 0.23 million wear cycles. Before this point, the spectral coherence is highest for carrier frequencies around 30–50 kHz, this being interpreted as being caused by the initial rough surface prepared using sandpaper, which exhibits high spatial frequency content. The initial high carrier frequency with high spectral coherence, therefore, supports the hypothesis of Section 4.2.2 that the gear mesh-modulated CS2 carrier frequency range is closely connected with the spatial frequency of the gear tooth surfaces.

After the run-in period (i.e. after the 0.23 million cycles), the roughening marks were worn away, and fatigue pitting started propagating. During this process, the dominant carrier frequency band jumps to the low-frequency range, below 15 kHz. This phenomenon matches the hypothesis (in Section 4.2.2) that sliding surfaces with low spatial frequency content, such as from fatigue pitting, will induce CS2 vibrations with lower carrier frequencies. Note that the distinctive vertical parallel bands shown in Figure 4.7 indicate the resonances of the gear system. This phenomenon will be further explained with the spectral coherence maps under different rotational speeds.

The ICS2-based selection results at 10 Hz are demonstrated in Figure 4.8. Similar to Figure 4.7, before 0.23 million cycles, a high carrier frequency range is selected. After 0.23 million cycles, the selected frequency band jumps to the low-frequency range. This again supports the hypothesis connecting surface spatial frequency and

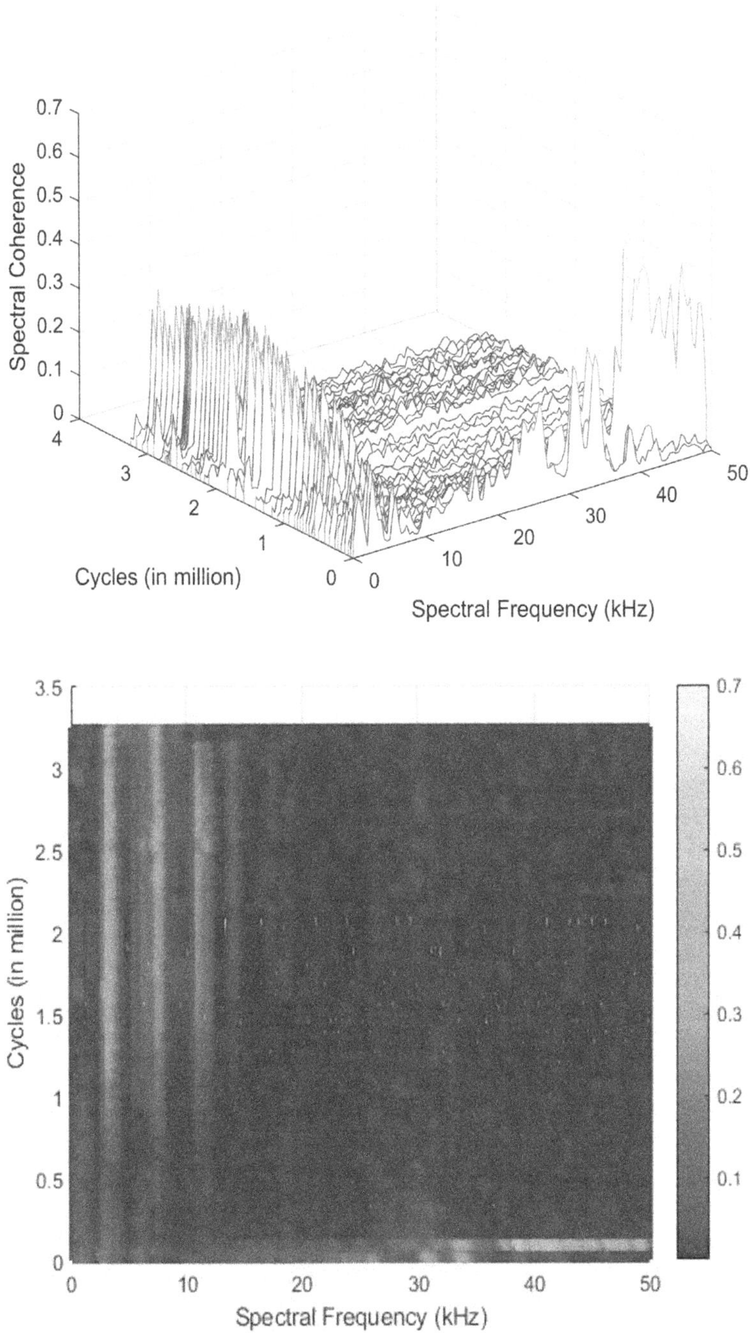

FIGURE 4.7 Lubricated test: Carrier frequency distribution at gear mesh (cyclic) frequency and top view. Made by the author [33].

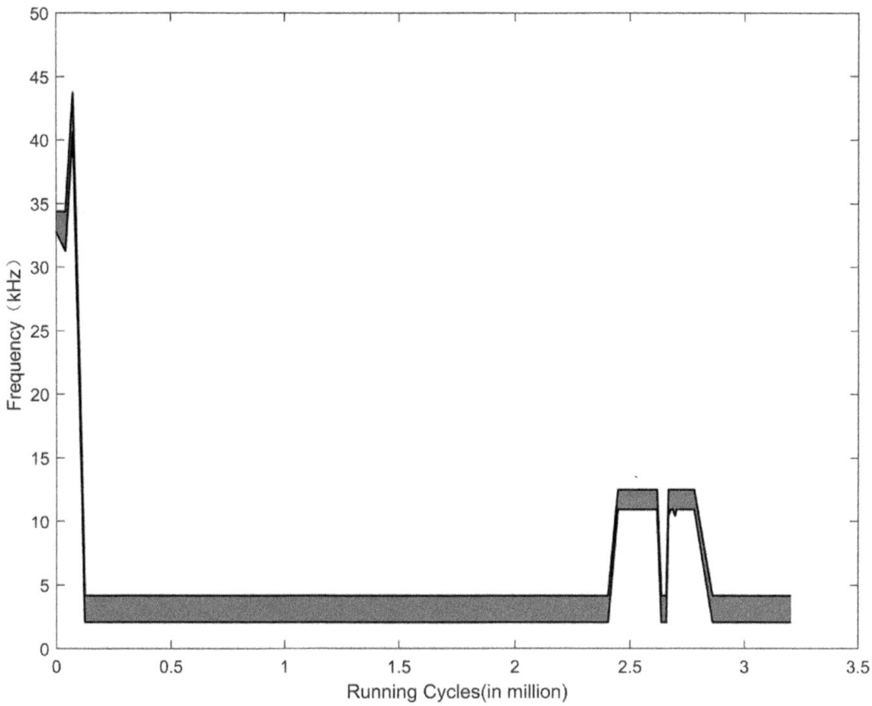

FIGURE 4.8 Lubricated test: ICS2-based band selection results. Made by the author [33].

CS2 carrier frequency, suggesting strong potential for differentiating between abrasive wear and fatigue pitting. Note that there is an abrupt change in [2.30~3.00] million cycles. During this period, there is a decrease in the total pitted area, which has been described in reference [30]. The cause of decreased pitted area might be that the generation of new pits achieves balance with the filling-in of existing cavities, perhaps from wear debris pressed into the cavities under the meshing load [48]. Thus, during this period, the excitation has changed, and so a different resonance becomes dominant in the response. The ICS2-based band selection results are affected, resulting in the occurrence of abrupt changes. After this 'abnormal' period, the ICS2-based band selection shifts back to the low frequency again, still indicating a low spatial frequency of pits. To help the reader obtain an intuitive understanding of this phenomenon, the gradual fatigue pitting propagation process is demonstrated in Figure 4.9.

To validate the fatigue pitting-related carrier frequency range, power spectral density (PSD) analysis [49–51] was applied to the scanned images (see Figure 4.10) to identify the spatial frequency affected by fatigue pitting initiation and propagation. Note that although the images in Figure 4.10 were captured at the same location of the same tooth, some features disappear due to new pits filling in existing cavities, resulting in some dissimilarities and a non-monotonic trend in the overall pitted area. However, the spatial frequencies represented during fatigue pitting propagation remain reliable. The PSD function provides a representation of the density of the squared amplitude (height

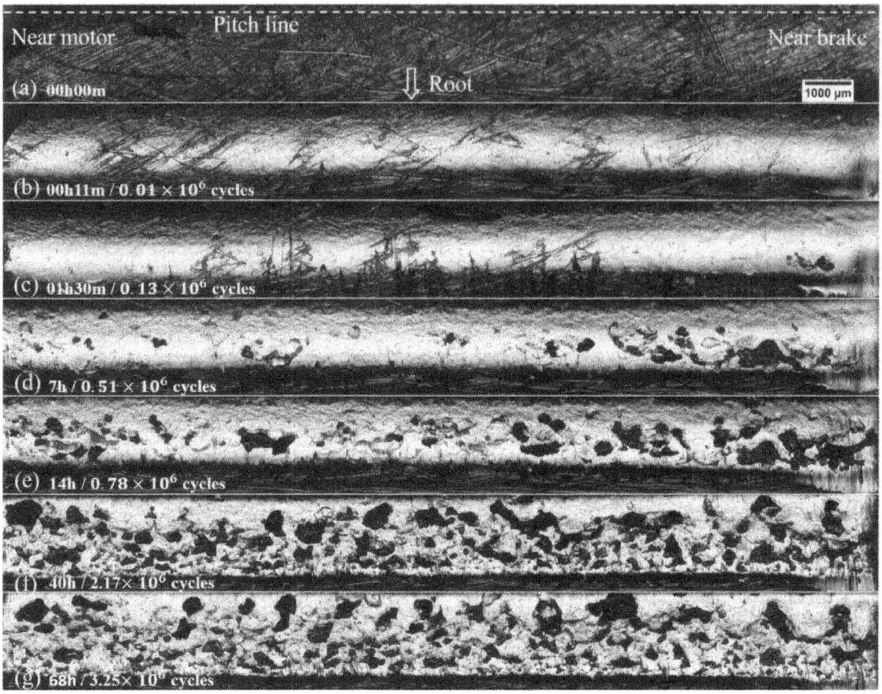

FIGURE 4.9 Figure pitting evolution on the dedendum of pinion tooth [30]. Photograph by H. Chang.

from a fixed reference) of a surface's morphology as a function of the spatial frequency of the morphology (units $\mu m^2 / \dfrac{1}{\mu m} = \mu m^3$). Spatial frequency is the inverse of the wavelength of the morphology features. The PSD analysis results are depicted in Figure 4.11, which shows that the spatial frequency range affected by fatigue pitting is [0.0012~0.0146] (1/μm). The sliding velocity (absolute value) range is [0~0.4461] (m/s) according to the dynamic model used in reference [21]. Based on Eq. (4.1), taking the central value of this sliding velocity range, the fatigue pitting-related vibration frequency range is expected to be around [0.27~3.27] kHz. The upper part of this range matches very closely the results shown in Figure 4.7 and Figure 4.8. The hypothesis for fatigue pitting from Section 4.2.2 – that fatigue pitting induces CS2 signals with low carrier frequencies – is therefore at least qualitatively supported by the lubricated test results. It is important to note that Eq. (4.1) is expected to be only qualitatively matching with the results, given the effect of transfer functions and background-noise distribution, which could bias the choice of the ICS2 indicator, for instance, towards an area where background noise is lower.

The relationship between the carrier frequency of the vibration and the spatial frequency of the gear surface has been investigated above, and the hypothesis has been supported. The effects of sliding velocity on the carrier frequency are investigated below.

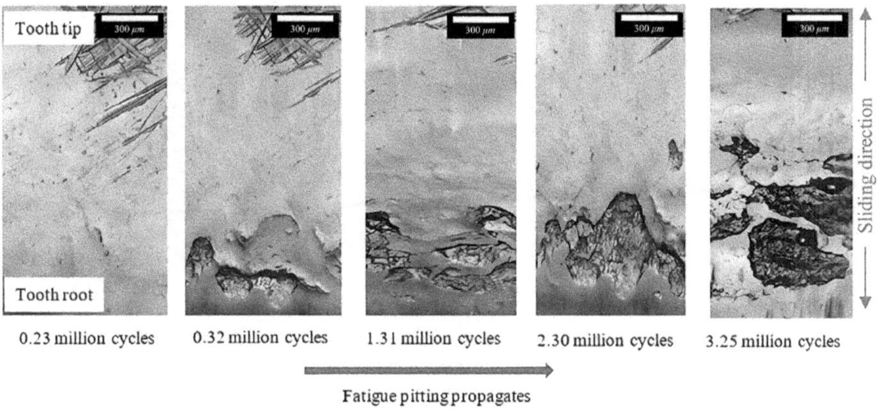

FIGURE 4.10 3D images of fatigue pitting propagation in the lubricated test. Photograph by the author [33].

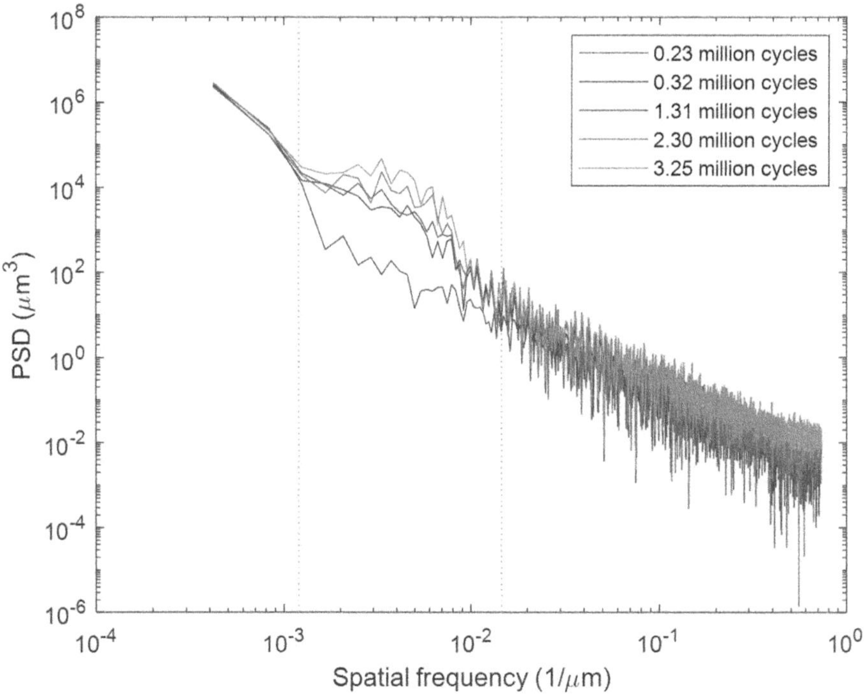

FIGURE 4.11 PSD analysis results of the scanned images of the lubricated test in Figure 4.10. Made by the author [33].

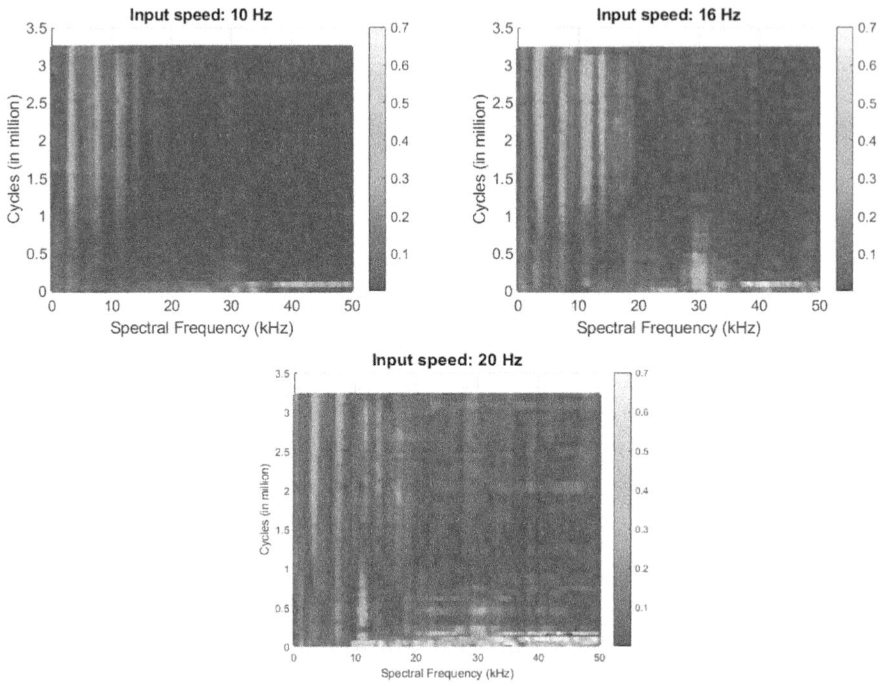

FIGURE 4.12 Carrier frequency distribution at gear mesh (cyclic) frequency at different speeds (top views) of the lubricated test. Made by the author [33].

Figure 4.12 shows the spectral coherence maps for different speeds: 10, 16, and 20 Hz. Next, a mean carrier frequency at gear mesh cyclic frequency was calculated for different speeds (Figure 4.13) over two frequency ranges. The mean carrier frequency was obtained based on the distribution of the spectral coherence along the spectral axis, for the cyclic component at gear mesh frequency. Figure 4.13(a) is averaged over the range 0–30 kHz, and Figure 4.13(b) over the range 0–50 kHz. Although Figure 4.12 shows that the dominant CS2 part (with the highest spectral coherence) is below 30 kHz at all speeds, the range of 0–50 kHz was included for unbiased comparison with the results, shown later, for dry wear (Figure 4.14). Figure 4.12 and Figure 4.13 show that an increase in sliding velocity (induced by increasing rotating speed) basically leads to an increase in mean carrier frequency, in accordance with Eq. (4.1). The fact that the calculated mean carrier frequency does not match with the raw product of the quantities on the right-hand side of Eq. (4.1) can be explained. It is largely due to background noise (most evident in the difference between Figure 4.13(a) and (b), with the uniform noise above 30 kHz giving a bias to a higher mean frequency, in particular for the 10 Hz result), and the use of spectral coherence rather than actual spectral power. The choice of spectral coherence, itself a measure of cyclostationarity, will bias the position of the dependent mean carrier frequency, but it was necessary to amplify the fairly weak CS2 content, in particular near modulated resonances, where

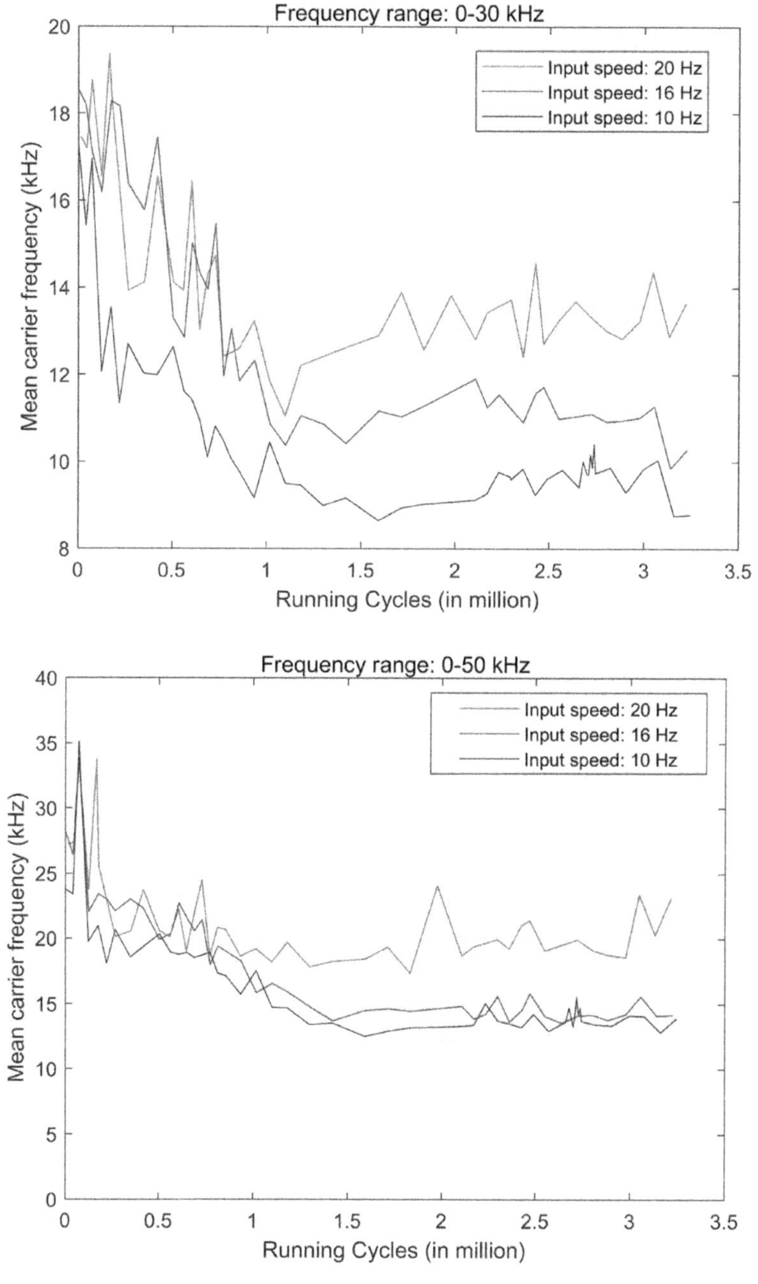

FIGURE 4.13 Mean carrier frequency with different rotating speeds of the lubricated test. Made by the author [33].

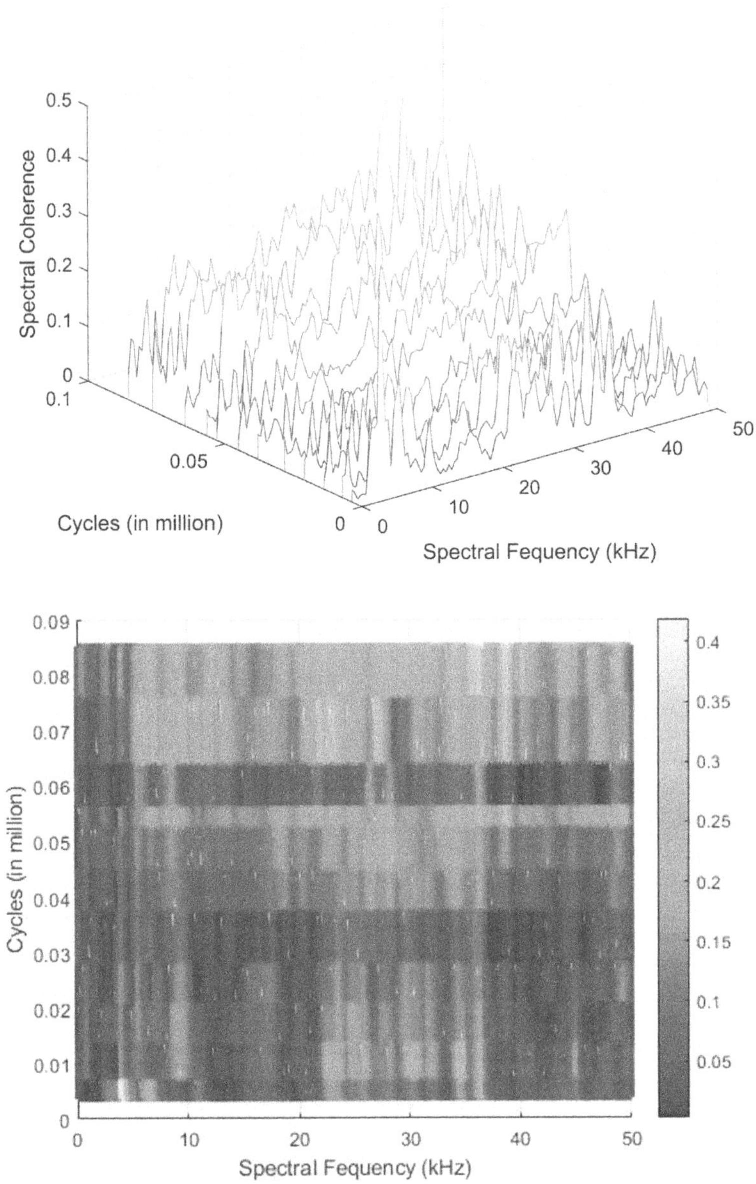

FIGURE 4.14 Dry test: Carrier frequency distribution at gear mesh (cyclic) frequency and top view. Made by the author [33].

the signal/noise ratio is higher, yet still limit the strong effects of resonances. Despite this limitation, the fixed resonances are clearly evident in Figure 4.12 at the different speeds, and rather than the mean frequency directly obeying Eq. (4.1), as for the expected *excitation* band, which would move gradually (on a continuum) depending on speed and spatial frequency, the dominant carrier frequency in the *response* would tend to move discretely between dominant bands. Nonetheless, the qualitative effect of sliding velocity on the vibration carrier frequency is clearly supported.

When ICS2-based band selection results, as shown in Figure 4.8, were compared for three different speeds, despite all indicating carrier frequencies within the range up to 20 kHz (as indicated in Figure 4.12, at least after the initial run-in period), the variation with speed was not monotonic, but this can be explained. The ICS2 value is much more sensitive to resonances than spectral coherence, and is also a sum over a number of harmonics of gear mesh frequency (rather than just the first, as for Figure 4.12). It is quite possible that at low speed, a high harmonic of the gear mesh frequency will be greatly amplified by a high-frequency resonance and bias the effective carrier frequency. The ICS2 parameters are more directly affected by the amount of modulation, and are used later (Section 4.4) as indicators of the severity of wear.

4.3.3.2 Observation Results of the Dry Test

Again, spectral coherence analysis is applied to the dry test to investigate the carrier frequency distribution with abrasive wear. The carrier frequency distribution (in terms of spectral coherence) at the gear mesh (cyclic frequency) for the dry test records is illustrated in Figure 4.14.

Unlike fatigue pitting, the abrasive wear propagation results in an increasing spectral coherence of the high carrier frequency part. It suggests that abrasion and the corresponding high spatial frequency surfaces generate higher carrier frequencies. This phenomenon is also confirmed by the ICS2-based band selection in Figure 4.15. Note that the initial surfaces were smooth in this test, and hence, a low carrier frequency was induced initially (unlike with the lubricated test). To verify the abrasive wear-related carrier frequency range, PSD analysis was applied to the scanned images (see Figure 4.16). Figure 4.17 shows the PSD analysis results, from which it can be found that the spatial frequency range affected by abrasive wear is [0.0283~0.0946] (1/μm). The sliding velocity (absolute value) range is around [0~0.4461] (m/s). Based on the right-hand side of Eq. (4.1), taking the average sliding velocity, the abrasive wear-related vibration frequency should be [6.31~21.10] kHz. It is in the high-frequency range. Figure 4.14 and Figure 4.15 show that the spectral coherence of the high carrier frequency part increases with an increase in abrasive wear, strongly supporting the hypothesis made in Section 4.2.2.

An important comment must be added regarding the different lubrication conditions in the two tests. These could have also affected the frequency range of the CS2 content, and future tests will be required to analyse the extent of this bias. However, the fact that the initial (healthy) cases for both dry and lubricated tests showed an opposite trend vs the corresponding worn cases is an initial indication that the bias is not likely to compromise the procedure. In fact, dry tests with smooth surfaces resulted in low-frequency vibrations, like the worn lubricated cases, and the initial lubricated tests (with the artificially roughened surface) were comparable to worn dry tests. Albeit based on a low number of observations, this supports the hypothesis that the wavelength of surface roughness strongly impacts the vibration spectral support, and that this effect seems fairly consistent with different lubrication conditions.

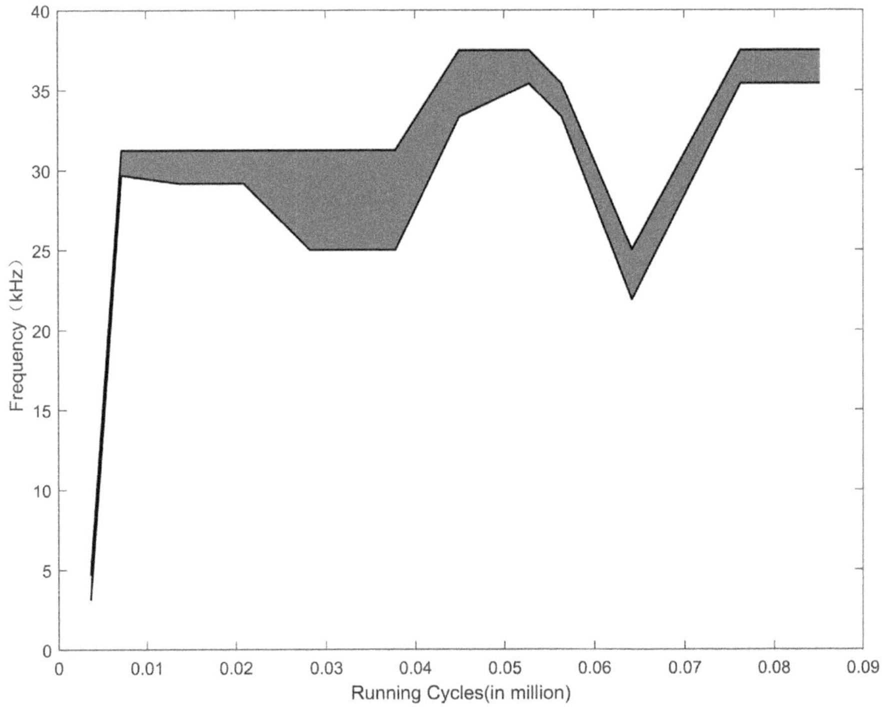

FIGURE 4.15 Dry test: ICS2-based band selection result. Made by the author [33].

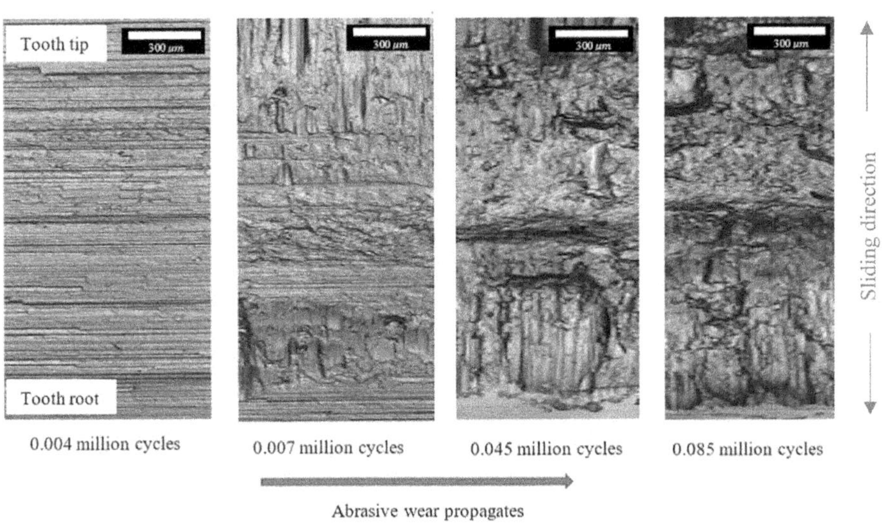

FIGURE 4.16 3D images of abrasive wear propagation in the dry test. Photograph by the author [33].

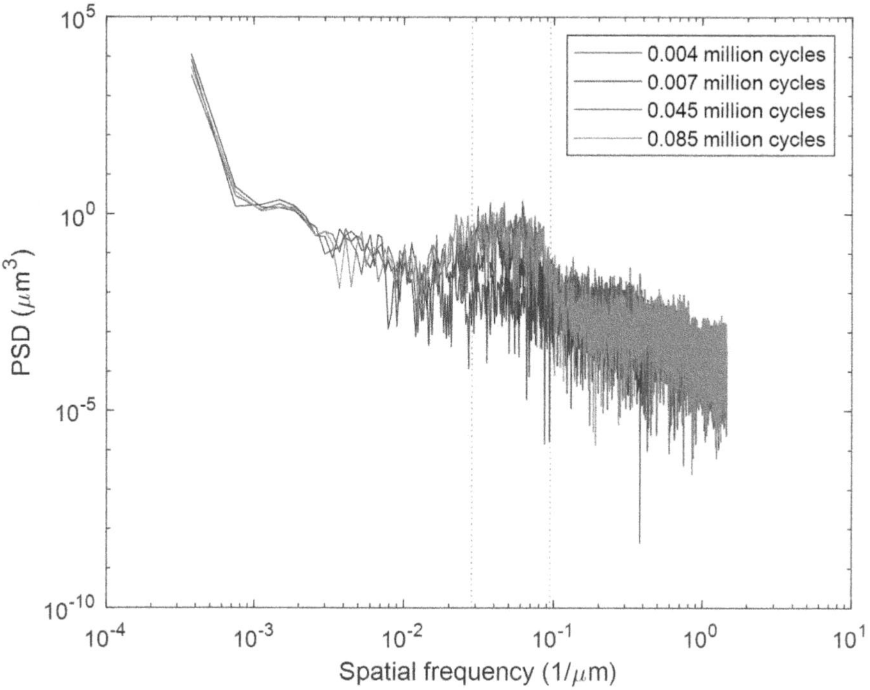

FIGURE 4.17 PSD analysis results of the 3D images shown in Figure 4.16. Made by the author [33].

4.4 A NEW VIBRATION-BASED PROCEDURE FOR COMPREHENSIVE GEAR WEAR MONITORING: MECHANISM IDENTIFICATION AND SEVERITY TRACKING

In Section 4.3, the hypothesis for fatigue pitting and abrasive wear identification was supported by the analysis of the spectral frequency distribution in the spectral coherence map and the ICS2-based band selection results. However, even though fatigue pitting and abrasive wear can be identified by observing the spectral coherence map and ICS2-based band selection results, the wear evolution (fatigue pitting and abrasive wear) cannot be monitored. A vibration-based approach/indicator is therefore still needed, which can distinguish the two wear mechanisms and track their evolution. A new vibration-based procedure for comprehensive gear wear monitoring is introduced and presented below.

The ability of ICS2 to monitor fatigue pitting and abrasive wear propagation is likely to be improved in the selected frequency band, which enhances CS2 content over background noise. Therefore, an ICS2-based approach can help comprehensively monitor gear wear progression: Wear mechanism identification and wear severity tracking. The basic procedure of the proposed vibration-based approach for comprehensive gear wear monitoring is illustrated in Figure 4.18.

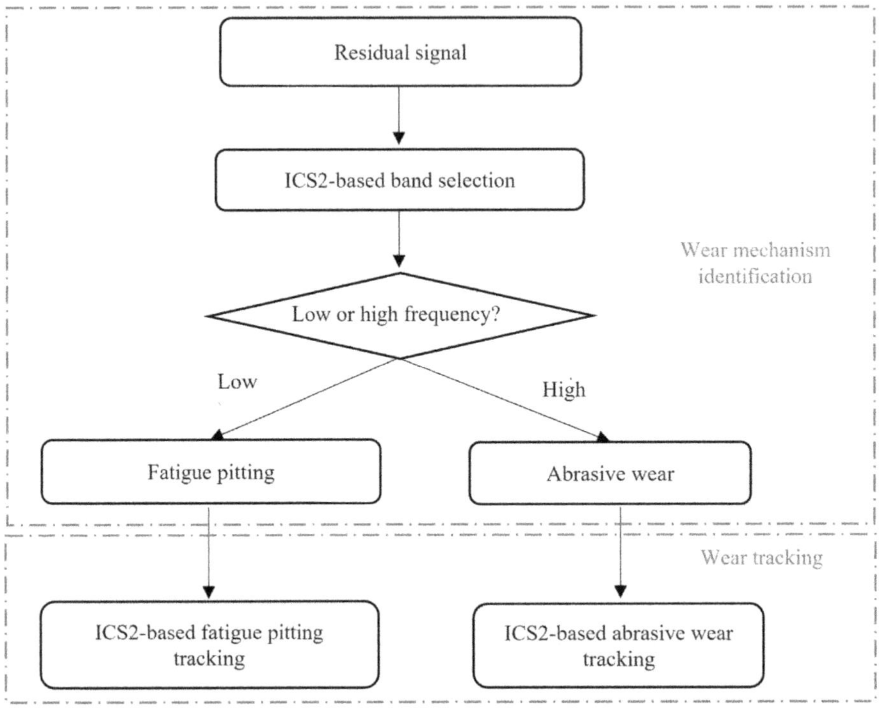

FIGURE 4.18 Guideline for vibration-based comprehensive gear wear monitoring. Made by the author [33].

The performance of this developed procedure for gear wear monitoring is demonstrated step by step, using experimental data from the designed lubricated and dry tests.

Firstly, ICS2-based band selection for identifying fatigue pitting ([2~13] kHz) and abrasive wear ([20~37] kHz) was illustrated in Figure 4.8 and Figure 4.15, demonstrating that fatigue pitting induces low carrier frequencies, while abrasive wear induces high carrier frequencies. Figure 4.19 shows that the ICS2 band selection results can assist the vibration-based wear mechanism identification and wear monitoring, by tracking the evolution of fatigue pitting and abrasive wear (at the micro-level), with a low-frequency range (2~13 kHz) and high-frequency range (20~37 kHz) indicators, respectively. From the comparison of Figure 4.19 (a) and (b), it is evident that only the ICS2 with low-frequency carrier tracks fatigue pitting propagation (in terms of pitted area). In contrast, in Figure 4.19 (c) and (d), only the ICS2 with a high-frequency carrier tracks the change in abrasive wear micro-scale features (in terms of surface roughness). Therefore, with an approximate frequency band (suggested by the ICS2-based band selection result), ICS2 can track wear severity for fatigue pitting and abrasive wear (at the micro-scale). Meanwhile, the performance of ICS2 in the low- and high-frequency ranges can also indicate different wear mechanisms. If ICS2 has a monotonic trend in the low-frequency range, but not in

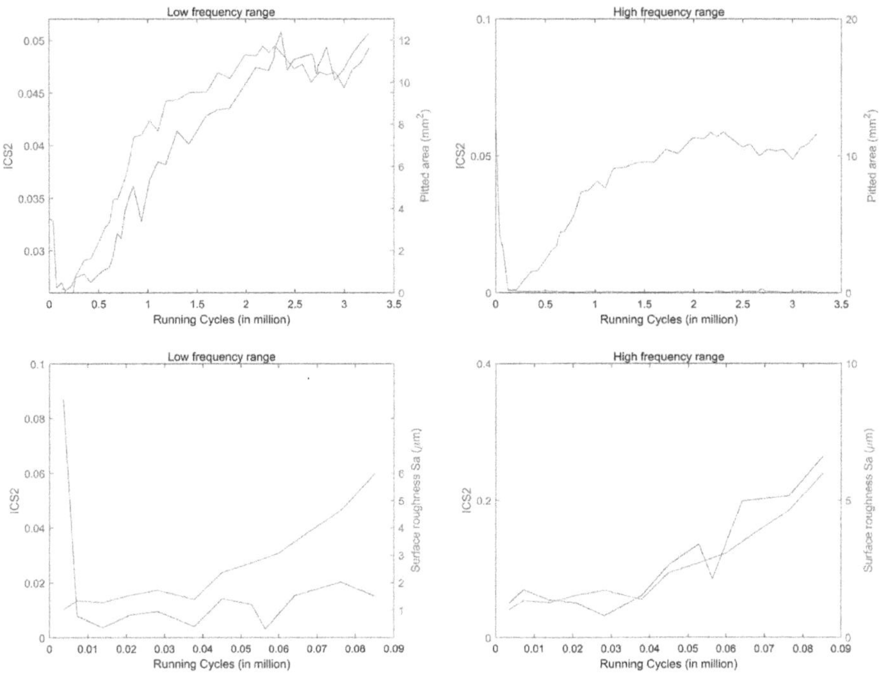

FIGURE 4.19 Performance of ICS2 in wear severity tracking. Made by the author [33].

the high-frequency range, it indicates fatigue pitting propagation, while a monotonic trend in the high-frequency range, but not in the low-frequency range, indicates abrasive wear propagation.

To quantify the performance of ICS2 in the low-frequency range in fatigue pitting severity tracking, the pitted area was used as a reference for correlation analysis. Correlation analysis was also applied comparatively on a number of other classical vibration indicators, such as RMS and kurtosis, to demonstrate the performance of band-selected ICS2 to monitor fatigue pitting progression, as shown in Figure 4.20.

Figure 4.21 shows that the ICS2 of vibration with the low-frequency carrier has a high correlation coefficient of 0.9085 with the total pitted area and can thus track the severity of fatigue pitting. The correlation analysis results for other classical indicators are summarised in Table 4.1. These show that ICS2 of vibration in the low-frequency range has the best performance in tracking the fatigue pitting propagation. Although the RMS value of the raw vibration signal also has a high correlation with the total pitted area, it mainly indicates the energy change of deterministic components of vibration, which have less physical relevance with fatigue pitting propagation (induced micro-level surface feature change). Notably, the amplitude of the 1st gear mesh harmonics of vibration has a low correlation coefficient with fatigue pitting propagation, supporting the earlier point that fatigue pitting has negligible effects on the gear tooth profile. Thus, this indicator should not have the

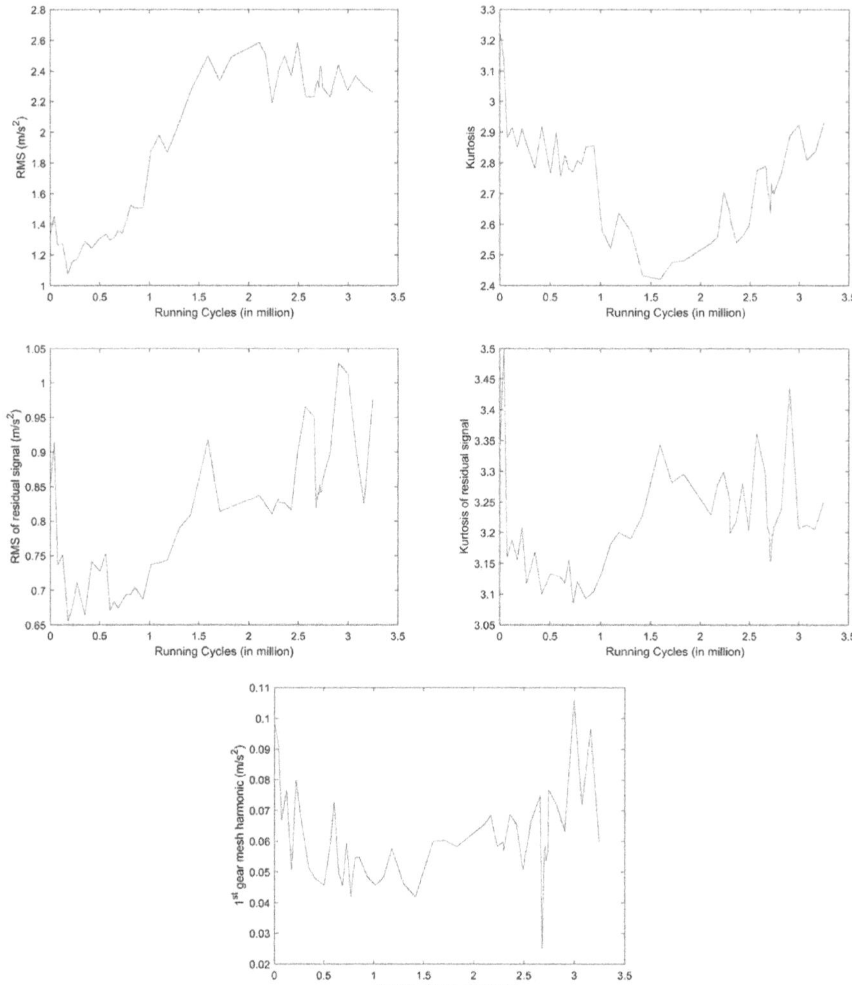

FIGURE 4.20 Classical indicators performance in the lubricated test. Made by the author [33].

ability to track fatigue pitting propagation. It should be pointed out that the sidebands were used in some references [52–54] to indicate damage, but they can only pick up non-uniform effects and are not useful for the more typical uniform wear case studied here. From the comparisons, ICS2 in the low-frequency range performs best in tracking fatigue pitting severity.

Similarly, surface roughness S_a was used as the tribological parameter to assess the effectiveness of ICS2 with the high-frequency carrier in monitoring abrasive wear propagation (at the micro-scale), as shown in Figure 4.22. The high correlation coefficient of 0.9086 shows that it can track this wear indicator very well. Note that compared with existing studies [24, 25], restricting the spectral band of the ICS2 to

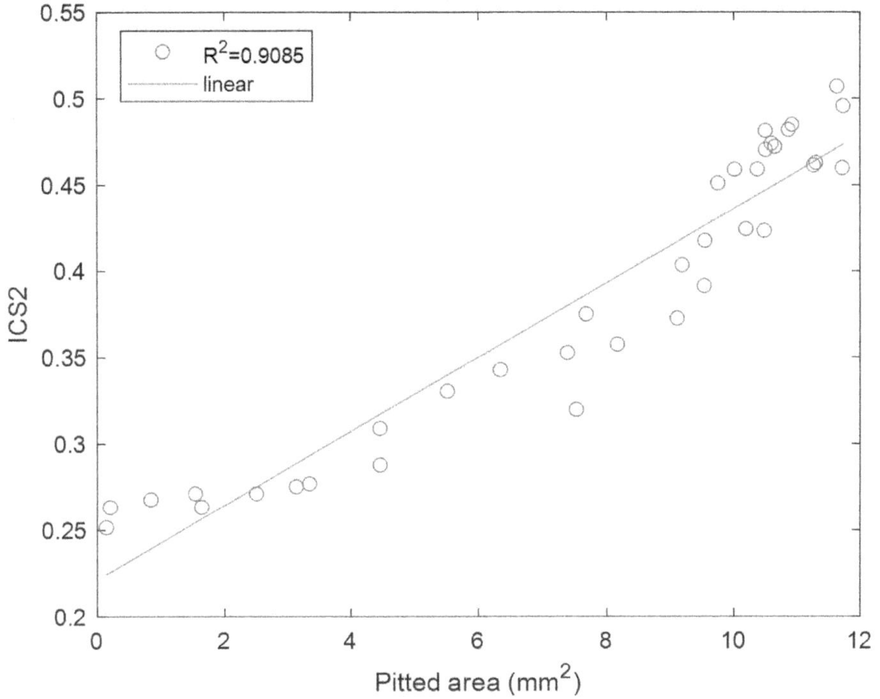

FIGURE 4.21 Lubricated test: Correlation between ICS2 and pitted area [33]. Made by the author.

TABLE 4.1
Correlation Analysis of Vibration Indicators with Tribological Parameters in the Lubricated Test

Indicators	Correlation with pitted area: R^2
ICS2 (high frequency range)	0.0215
ICS2 (low frequency range)	0.9085
RMS (raw signal)	0.8146
Kurtosis (raw signal)	0.1237
RMS (residual signal)	0.6984
Kurtosis (residual signal)	0.6718
1st gear mesh frequency amplitude	0.2476

the high-frequency range, instead of the full frequency band, makes ICS2 a much better index for tracking surface roughness changes.

Compared with other classical indicators (in Figure 4.23), ICS2 also has the best performance in monitoring the abrasive wear-induced micro-level surface feature changes, as shown in Table 4.2. Although the RMS value of the raw signal has a low

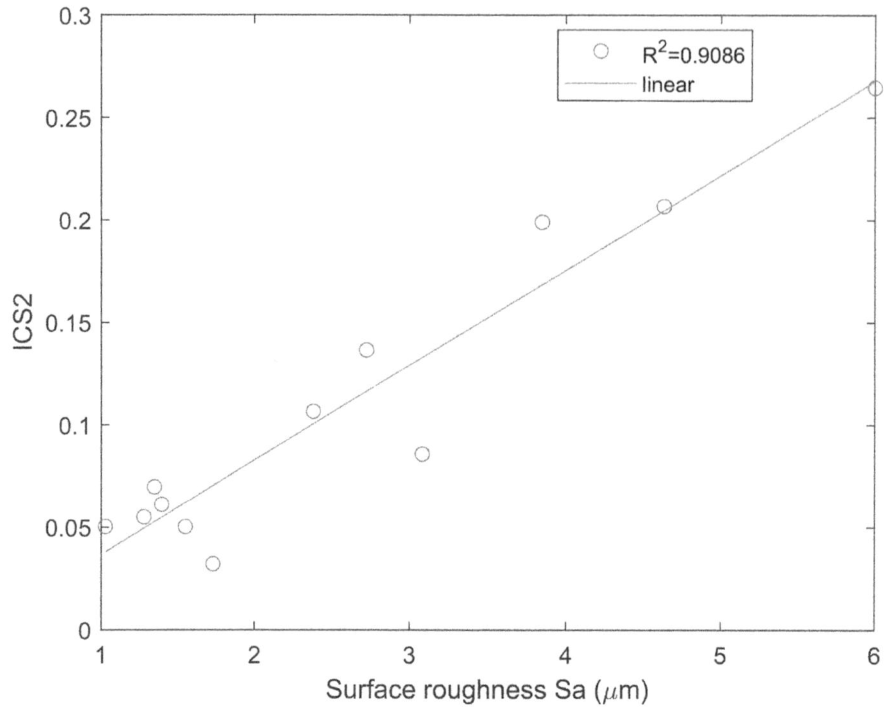

FIGURE 4.22 Dry test: Correlation between ICS2 and surface roughness. Made by the author [33].

correlation with the surface roughness, it should have a strong relationship with tooth profile changes induced by abrasive wear, which is shown in Figure 4.24. However, while Figure 4.24 shows that the RMS value has a high correlation ($R^2 = 0.9816$) with the wear depth before 0.045 million cycles, it deviates widely after that. In the light of the findings of reference [55], it seems that the vibrations only respond to deviations of the tooth profile around the mean wear, and measurement of 'absolute TE' is required to determine the overall wear depth. Further investigation will be conducted in future work. However, the total wear in this test was extreme, and that corresponding to 0.045 million cycles here would often be considered as being at the limit.

4.5 SUMMARY

With consideration of the underlying physics of the gear meshing process and the unique surface features induced by fatigue pitting and abrasive wear, this chapter has investigated vibration-based methods to identify these two wear mechanisms and then track their evolution. This development is based on the cyclostationary analysis technique, which is applied for the first time to analyse wear-related low-energy phenomena (friction, asperity contacts) in vibration signals. Different from existing studies for

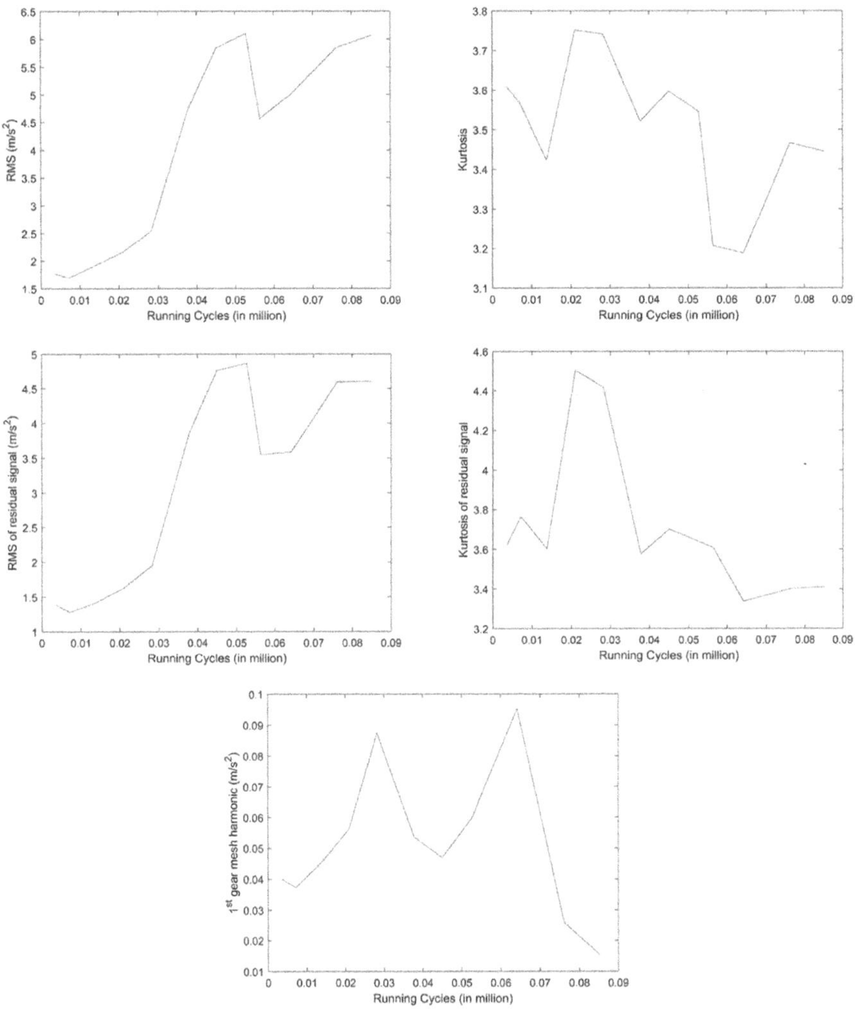

FIGURE 4.23 Classical indicators performance in the dry test. Made by the author [33].

wear mechanism identification (such as analysing wear particles or images captured from gear tooth surface), the proposed approach can be implemented online, making it more efficient than wear debris analysis techniques. In the presented method, an indicator of second-order cyclostationarity of the vibration signal, ICS2, is calculated for low and high carrier spectral frequencies and then used to separate fatigue pitting and abrasive wear. Moreover, the use of specific spectral bands for the calculation of the ICS2 increases the indicator's capability to track the evolution of fatigue pitting and abrasive wear (micro-level). As discussed in Section 4.1, none of the existing vibration-based gear wear monitoring research involved two wear mechanisms/phenomena identification and monitoring. The ICS2-based wear monitoring results

TABLE 4.2
Correlation Analysis of Vibration Indicators with the Surface Roughness in the Dry Test

Indicators	Correlation with surface roughness: R^2
ICS2 (high frequency range)	0.9086
ICS2 (low frequency range)	0.0240
RMS of the raw signal	0.5590
Kurtosis of the raw signal	0.2313
RMS of the residual signal	0.4895
Kurtosis of the residual signal	0.2611
1st gear mesh frequency amplitude	0.0661

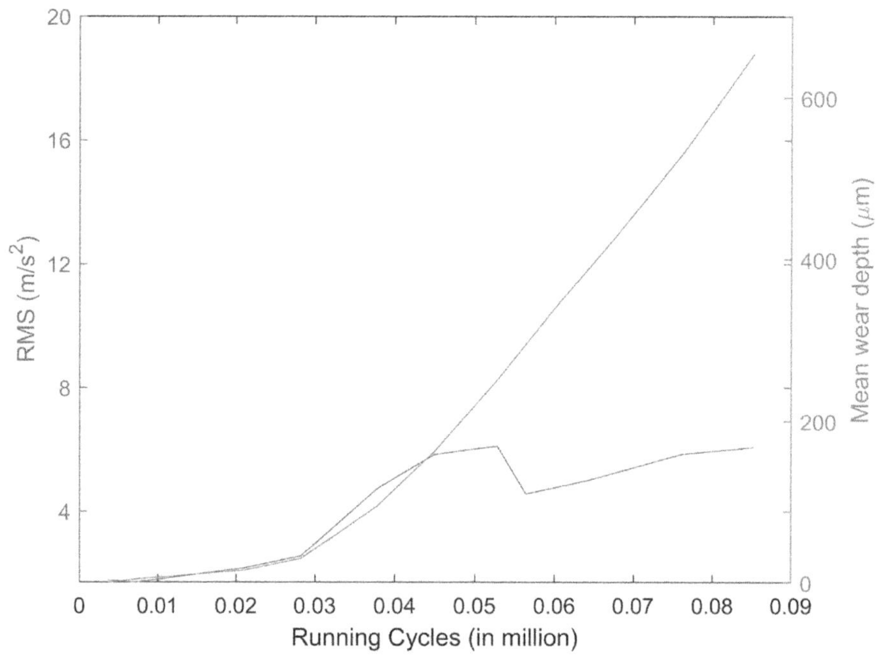

FIGURE 4.24 RMS performance of raw signal in tracking wear depth change (from wear particles). Made by the author [33].

can provide useful information for the monitoring and prognostics of gear systems. Experimental data support the effectiveness of the proposed vibration-based method.

However, it should be noted that the spectral frequency distribution of measured vibrations can be affected by lots of factors, such as load, speed, and especially lubrication. Therefore, further investigation into the impact of operational conditions on the distribution of spectral frequencies is recommended.

REFERENCES

[1] P.J.L. Fernandes, C. McDuling, Surface contact fatigue failures in gears, *Engineering Failure Analysis*, 4 (1997) 99–107. https://doi.org/10.1016/S1350-6307(97)00006-X.

[2] S. Glodež, Z. Ren, G. Fajdiga, Computational modelling of the surface fatigue crack growth on gear teeth flanks, *Communications in Numerical Methods in Engineering*, 17 (2001) 529–541. https://doi.org/10.1002/cnm.427.

[3] J.A. Williams, A.M. Hyncica, Mechanisms of abrasive wear in lubricated contacts, *Wear*, 152 (1992) 57–74. https://doi.org/10.1016/0043-1648(92)90204-L.

[4] American Gear Manufacturers Association, *Appearance of gear teeth: Terminology of wear and failure*, Vol 40, American Gear Manufacturers Association, Virginia, United States, 1995.

[5] R.B. Randall, *Vibration-based condition monitoring: industrial, aerospace and automotive applications*, Wiley, Hoboken, United States, 2011.

[6] Q. Ni, J.C. Ji, K. Feng, B. Halkon, A novel correntropy-based band selection method for the fault diagnosis of bearings under fault-irrelevant impulsive and cyclostationary interferences, *Mechanical Systems and Signal Processing*, 153 (2021) 107498. https://doi.org/10.1016/j.ymssp.2020.107498.

[7] Y.-F. Li, M. Zuo, K. Feng, Y.-J. Chen, Detection of bearing faults using a novel adaptive morphological update lifting wavelet, *Chinese Journal of Mechanical Engineering*, 30 (2017) 1305–1313. https://doi.org/10.1007/s10033-017-0186-1.

[8] Q. Ni, K. Feng, K. Wang, B. Yang, Y. Wang, A case study of sample entropy analysis to the fault detection of bearing in wind turbine, *Case Studies in Engineering Failure Analysis*, 9 (2017) 99–111. https://doi.org/10.1016/j.csefa.2017.10.002.

[9] K. Feng, K. Wang, P. Chen, *A signal selection scheme through order tracking techniques for planetary gearbox condition monitoring*, 2016 Prognostics and System Health Management Conference (PHM-Chengdu), IEEE, 2016, pp. 1–5.

[10] T. Barszcz, R.B. Randall, Application of spectral kurtosis for detection of a tooth crack in the planetary gear of a wind turbine, *Mechanical Systems and Signal Processing*, 23 (2009) 1352–1365. https://doi.org/10.1016/j.ymssp.2008.07.019.

[11] K. Feng, K. Wang, M. Zhang, Q. Ni, M.J. Zuo, A diagnostic signal selection scheme for planetary gearbox vibration monitoring under non-stationary operational conditions, *Measurement Science and Technology*, 28 (2017) 035003. https://doi.org/10.1088/1361-6501/aa543e.

[12] H. Endo, R.B. Randall, C. Gosselin, Differential diagnosis of spall vs. cracks in the gear tooth fillet region: Experimental validation, *Mechanical Systems and Signal Processing*, 23 (2009) 636–651. https://doi.org/10.1016/j.ymssp.2008.08.015.

[13] D. Peng, W.A. Smith, P. Borghesani, R.B. Randall, Z. Peng, Comprehensive planet gear diagnostics: Use of transmission error and mesh phasing to distinguish localised fault types and identify faulty gears, *Mechanical Systems and Signal Processing*, 127 (2019) 531–550. https://doi.org/10.1016/j.ymssp.2019.03.024.

[14] R. Ma, Y. Chen, Research on the dynamic mechanism of the gear system with local crack and spalling failure, *Engineering Failure Analysis*, 26 (2012) 12–20. https://doi.org/10.1016/j.engfailanal.2012.05.022.

[15] Y. Li, K. Ding, G. He, H. Lin, Vibration mechanisms of spur gear pair in healthy and fault states, *Mechanical Systems and Signal Processing*, 81 (2016) 183–201. https://doi.org/10.1016/j.ymssp.2016.03.014.

[16] M.J. Roan, J.G. Erling, L.H. Sibul, A new, non-linear, adaptive, blind source separation approach to gear tooth failure detection and analysis, *Mechanical Systems and Signal Processing*, 16 (2002) 719–740. https://doi.org/10.1006/mssp.2002.1504.

[17] K. Feng, K. Wang, Q. Ni, M.J. Zuo, D. Wei, A phase angle based diagnostic scheme to planetary gear faults diagnostics under non-stationary operational conditions, *Journal of Sound and Vibration*, 408 (2017) 190–209. https://doi.org/10.1016/j.jsv.2017.07.030.

[18] C. Kar, A.R. Mohanty, Monitoring gear vibrations through motor current signature analysis and wavelet transform, *Mechanical Systems and Signal Processing*, 20 (2006) 158–187. https://doi.org/10.1016/j.ymssp.2004.07.006.

[19] P.V. Kane, A.B. Andhare, Critical evaluation and comparison of psychoacoustics, acoustics and vibration features for gear fault correlation and classification, *Measurement*, 154 (2020) 107495. https://doi.org/10.1016/j.measurement.2020.107495.

[20] H. Ding, A. Kahraman, Interactions between nonlinear spur gear dynamics and surface wear, *Journal of Sound and Vibration*, 307 (2007) 662–679. https://doi.org/10.1016/j.jsv.2007.06.030.

[21] K. Feng, P. Borghesani, W.A. Smith, R.B. Randall, Z.Y. Chin, J. Ren, Z. Peng, Vibration-based updating of wear prediction for spur gears, *Wear*, 426–427 (2019) 1410–1415. https://doi.org/10.1016/j.wear.2019.01.017.

[22] A. Soom, J.-W. Chen, Simulation of random surface roughness-induced contact vibrations at hertzian contacts during steady sliding, *Journal of Tribology*, 108 (1986) 123–127. https://doi.org/10.1115/1.3261131.

[23] S. Li, A. Kahraman, A tribo-dynamic model of a spur gear pair, *Journal of Sound and Vibration*, 332 (2013) 4963–4978. https://doi.org/10.1016/j.jsv.2013.04.022.

[24] X. Zhang, W.A. Smith, P. Borghesani, Z. Peng, R.B. Randall, *Use of cyclostationarity to detect changes in gear surface roughness using vibration measurements, asset intelligence through integration and interoperability and contemporary vibration engineering technologies*, Springer, Cham, Switzerland, 2019, pp. 763–771.

[25] Y. Yang, W.A. Smith, P. Borghesani, Z. Peng, R.B. Randall, Detecting changes in gear surface roughness using vibration signals, in *Proceedings of the Acoustics Conference*, Hunter Valley, NSW, Australia, 2015, pp. 15–18.

[26] R.F. Handschuh, A.J. Zakrajsek, High-pressure angle gears: Comparison to typical gear designs, *Journal of Mechanical Design*, 133(11) (2011) 114501. https://doi.org/10.1115/1.4004458.

[27] Z.K. Zhu, Z.H. Feng, F.R. Kong, Cyclostationarity analysis for gearbox condition monitoring: Approaches and effectiveness, *Mechanical Systems and Signal Processing*, 19 (2005) 467–482. https://doi.org/10.1016/j.ymssp.2004.02.007.

[28] C. Capdessus, M. Sidahmed, J.L. Lacoume, Cyclostationary processes: Application in gear faults early diagnosis, *Mechanical Systems and Signal Processing*, 14 (2000) 371–385. https://doi.org/10.1006/mssp.1999.1260.

[29] R. Zimroz, W. Bartelmus, Gearbox condition estimation using cyclo-stationary properties of vibration signal, *Key Engineering Materials*, 413–414 (2009) 471–478. https://doi.org/10.4028/www.scientific.net/KEM.413-414.471.

[30] H. Chang, P. Borghesani, W.A. Smith, Z. Peng, Application of surface replication combined with image analysis to investigate wear evolution on gear teeth – A case study, *Wear*, 430–431 (2019) 355–368. https://doi.org/10.1016/j.wear.2019.05.024.

[31] I. Barányi, G. Kalácska, P. De Baets, Characterisation of abrasive worn surfaces by surface microtopography parameters, *International Journal Sustainable Construction & Design*, 4(2) (2013). https://doi.org/10.21825/scad.v4i2.1036.

[32] B. Bhushan, *Modern tribology handbook, two volume set*, CRC press, Boca Raton, United States, 2000.

[33] K. Feng, W.A. Smith, P. Borghesani, R.B. Randall, Z. Peng. Use of cyclostationary properties of vibration signals to identify gear wear mechanisms and track wear evolution, *Mechanical Systems and Signal Processing*, 150 (2021) 107258. https://doi.org/10.1016/j.ymssp.2020.107258.

[34] P. Velex, V. Cahouet, Experimental and numerical investigations on the influence of tooth friction in spur and helical gear dynamics, *Journal of Mechanical Design*, 122 (2000) 515–522. https://doi.org/10.1115/1.1320821.

[35] J. Antoni, G. Xin, N. Hamzaoui, Fast computation of the spectral correlation, *Mechanical Systems and Signal Processing*, 92 (2017) 248–277. https://doi.org/10.1016/j.ymssp.2017.01.011.

[36] R.W. Bruce, *Handbook of lubrication and tribology, volume II: Theory and design*, CRC Press, Boca Raton, United States, 2012.

[37] F. Combet, L. Gelman, Optimal filtering of gear signals for early damage detection based on the spectral kurtosis, *Mechanical Systems and Signal Processing*, 23 (2009) 652–668. https://doi.org/10.1016/j.ymssp.2008.08.002.

[38] Y. Chen, X. Liang, M.J. Zuo, Sparse time series modeling of the baseline vibration from a gearbox under time-varying speed condition, *Mechanical Systems and Signal Processing*, 134 (2019) 106342. https://doi.org/10.1016/j.ymssp.2019.106342.

[39] J.M. Ha, B.D. Youn, H. Oh, B. Han, Y. Jung, J. Park, Autocorrelation-based time synchronous averaging for condition monitoring of planetary gearboxes in wind turbines, *Mechanical Systems and Signal Processing*, 70–71 (2016) 161–175. https://doi.org/10.1016/j.ymssp.2015.09.040.

[40] S. Braun, The synchronous (time domain) average revisited, *Mechanical Systems and Signal Processing*, 25 (2011) 1087–1102. https://doi.org/10.1016/j.ymssp.2010.07.016.

[41] J. Antoni, R.B. Randall, Unsupervised noise cancellation for vibration signals: Part II – A novel frequency-domain algorithm, *Mechanical Systems and Signal Processing*, 18 (2004) 103–117. https://doi.org/10.1016/S0888-3270(03)00013-X.

[42] J. Antoni, F. Bonnardot, A. Raad, M. El Badaoui, Cyclostationary modelling of rotating machine vibration signals, *Mechanical Systems and Signal Processing*, 18 (2004) 1285–1314. https://doi.org/10.1016/S0888-3270(03)00088-8.

[43] Z. Feng, M.J. Zuo, R. Hao, F. Chu, M.E. Badaoui, Gear damage assessment based on cyclic spectral analysis, *IEEE Transactions on Reliability*, 60 (2011) 21–32. https://doi.org/10.1109/TR.2010.2104017.

[44] A. Raad, J. Antoni, M. Sidahmed, Indicators of cyclostationarity: Theory and application to gear fault monitoring, *Mechanical Systems and Signal Processing*, 22 (2008) 574–587. https://doi.org/10.1016/j.ymssp.2007.09.011.

[45] W.A. Smith, P. Borghesani, Q. Ni, K. Wang, Z. Peng, Optimal demodulation-band selection for envelope-based diagnostics: A comparative study of traditional and novel tools, *Mechanical Systems and Signal Processing*, 134 (2019) 106303. https://doi.org/10.1016/j.ymssp.2019.106303.

[46] Q. Ni, K. Wang, J. Zheng, Rolling element bearings fault diagnosis based on a novel optimal frequency band selection scheme, *IEEE Access*, 7 (2019) 80748–80766. https://doi.org/10.1109/ACCESS.2019.2921409.

[47] J. Antoni (2021). Fast_SC(x,Nw,alpha_max,Fs,opt) (www.mathworks.com/matlabcentral/fileexchange/60561-fast_sc-x-nw-alpha_max-fs-opt), MATLAB Central File Exchange. Retrieved June 14, 2021.

[48] P. Feng, P. Borghesani, H. Chang, W.A. Smith, R.B. Randall, Z. Peng, Monitoring gear surface degradation using cyclostationarity of acoustic emission, *Mechanical*

Systems and Signal Processing, 131 (2019) 199–221. https://doi.org/10.1016/j.ymssp.2019.05.055.

[49] J. Michalski, Surface topography of the cylindrical gear tooth flanks after machining, *The International Journal of Advanced Manufacturing Technology*, 43 (2008) 513. https://doi.org/10.1007/s00170-008-1737-5.

[50] G.M. Krolczyk, R.W. Maruda, P. Nieslony, M. Wieczorowski, Surface morphology analysis of duplex stainless steel (DSS) in clean production using the power spectral density, *Measurement*, 94 (2016) 464–470. https://doi.org/10.1016/j.measurement.2016.08.023.

[51] J. Michalski, P. Pawlus, W. Żelasko, Surface topography of cylindrical gear wheels after smoothing in abrasive mass, honing and shot peening, *Journal of Physics: Conference Series*, 311 (2011) 012022. https://doi.org/10.1088/1742-6596/311/1/012022.

[52] C. Hu, W.A. Smith, R.B. Randall, Z. Peng, Development of a gear vibration indicator and its application in gear wear monitoring, *Mechanical Systems and Signal Processing*, 76–77 (2016) 319–336. https://doi.org/10.1016/j.ymssp.2016.01.018.

[53] R.B. Randall, A new method of modeling gear faults, *Journal of Mechanical Design*, 104 (1982) 259–267. https://doi.org/10.1115/1.3256334.

[54] S. Ziaran, R. Darula, Determination of the state of wear of high contact ratio gear sets by means of spectrum and cepstrum analysis, *Journal of Vibration and Acoustics*, 135(2) (2013) 021008. https://doi.org/10.1115/1.4023208.

[55] Z.Y. Chin, W.A. Smith, P. Borghesani, R.B. Randall, Z. Peng, Absolute transmission error: A simple new tool for assessing gear wear, *Mechanical Systems and Signal Processing*, 146 (2021) 107070. https://doi.org/10.1016/j.ymssp.2020.107070

5 Dynamic Model Development

5.1 INTRODUCTION

The dynamic model, incorporating key properties such as gear meshing stiffness, transmission error, and damping, can simulate and represent gear system responses under different failure modes and severities. Its time and cost efficiency can bring significant benefits to gear wear analysis. As an output of the dynamic model, the dynamic contact force is a critical parameter and a crucial input to the tribological/wear model for wear analysis. Therefore, modelling the contact force is vital for analysing gear wear characteristics and further predicting its propagation.

From the literature review of Section 2.4.1, various approaches, such as the finite element model (FEM) [1–3] and empirical equations [4–9], have been developed/applied to estimate contact force, which is used as an input into tribological (wear) models. However, there is a common drawback existing in both FEM (without extra efforts in defining boundary conditions and mesh generations) and empirical equations, that is, only the contact force under quasi-static conditions can be produced and represented. However, in engineering practices, the gear transmission system is usually operated under dynamic operating conditions, and the corresponding responses are quite different from those under quasi-static conditions. Due to the inertia effects, the dynamic meshing forces are typically larger than the corresponding quasi-static forces, and their magnitudes and waveforms are quite different [10]. Therefore, to guarantee reliable wear analysis and prediction through tribological (wear) models, the dynamic contact force with inertia effects should be properly evaluated; thus, a 21-degree-of-freedoms (21-DOFs) dynamic model is established in this book based on the University of New South Wales (UNSW) gearbox test bench. To guarantee the outputs (such as dynamic contact forces and vibrations) from the dynamic model are close enough to the measurements from the experimental rig, the developed dynamic model has been validated and calibrated through a series of tests, including impact tests, speed ramp tests, and several constant speed tests. The development of this comprehensive dynamic model will be introduced with details in the following. It should be pointed out that the 21-DOFs lumped parameter dynamic model developed in this project is based on the fixed-axis spur gearbox test rig shown in Figure 3.2, whose structure and properties are different from the previously published research [11–13]. Also, as introduced in Chapter 3, the developed dynamic model will be integrated

DOI: 10.1201/9781003686736-5

with tribological models that can be updated according to vibrations to predict gear wear propagation. This overall integration architecture is the main novelty of this book, instead of the dynamic model development itself.

5.2 DYNAMIC MODEL STRUCTURE

The layout of the UNSW spur gearbox test rig and each labelled modelling component (such as coupling, motor, and brake) is shown in Figure 5.1. The major parameters that are included in the dynamic model are summarised in Table 5.1. A 21-DOFs lumped parameter dynamic model is developed based on the spur gearbox test rig (shown in Figure 5.1), whose skeleton is shown in Figure 5.2.

There is a total of 21 DOFs of this dynamic model, including nine torsional DOFs and 12 translational DOFs. The mass, stiffness, damping, and inertia parameter information of this developed dynamic model are summarised in Table 5.2 and Table 5.3. The basic motion equations describing the coupled torsional and translational model are described as:

$$M\ddot{x} + C\dot{x} + Kx = f \tag{5.1}$$

where

$$x = \left[\theta_1, \theta_2, \theta_3, \theta_5, \theta_7, \theta_8, \theta_{10}, \theta_{12}, \theta_{13}, y_4, y_5, y_6, y_9, y_{10}, y_{11}, x_4, x_5, x_6, x_9, x_{10}, x_{11}\right]^T \tag{5.2}$$

represents the translational (x_i, y_i) and angular displacements θ_i of the different nodes of the system in the plane perpendicular to the shaft axes, and $\boldsymbol{C}$, $\boldsymbol{K}$, and $\boldsymbol{f}$ are the corresponding damping, stiffness, and force matrices. The absence of bearing torsional DOFs ($\theta_4, \theta_6, \theta_9, \theta_{11}$) is due to the lack of torsional stiffness at the bearings, whereas the missing translational displacement DOFs (all the nodes outside the

TABLE 5.1
Basic Parameters of the Spur Gearbox Test Rig

Parameters	Pinion	Gear
Gear type	**Standard involute**	**Standard involute**
Modulus of elasticity, $\boldsymbol{E}$ (GPa)	205	205
Poisson's ratio, $\boldsymbol{\nu}$	0.29	0.29
Face width, $\boldsymbol{W}$ (mm)	20	20
Module, $\boldsymbol{m}$	2	2
Pressure angle, φ (deg)	20	20
Addendum (mm)	1.00	1.00
Dedendum (mm)	1.25	1.25
Number of teeth	19	20
Pitch radius, $\boldsymbol{r}$ (mm)	19	20

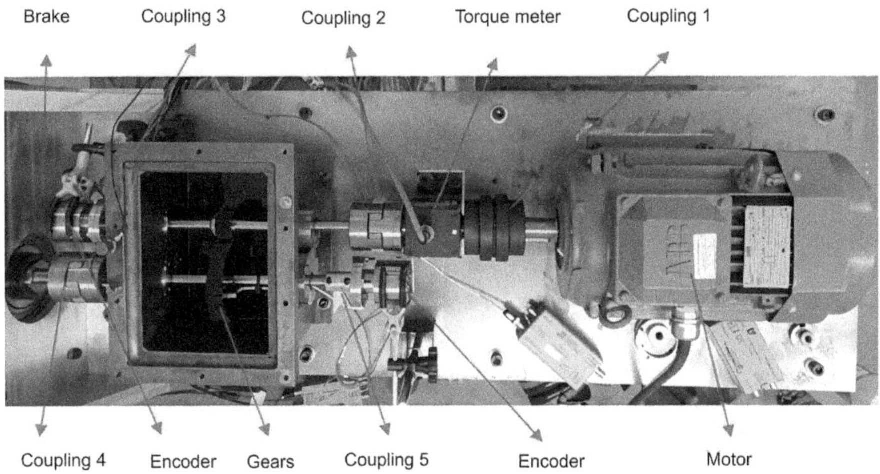

FIGURE 5.1 Spur gearbox test rig. Photograph by the author [14].

gearbox) is a result of the low bending stiffness of the joints, effectively isolating the linear displacements of the gearbox.

The force vector $\boldsymbol{f}$ includes the input and output torques $\boldsymbol{T}_{mot}$ and $\boldsymbol{T}_{brk}$ provided by the motor and brake (on θ_1 and θ_{13}, respectively), and the contact forces between the two gears, modelled as F_k and F_c (elastic and viscous components). These are simulated by combining a gear meshing stiffness k_m, damping coefficient c_m and geometric transmission error (GTE) e_t:

$$\mathrm{F_k} = k_m\left(R_{b1}\theta_5 - R_{b2}\theta_{10} - y_5 cos\varphi + y_{10} cos\varphi + x_5 sin\varphi - x_{10} sin\varphi + e_t\right) \tag{5.3}$$

$$\mathrm{F_c} = C_m\left(R_{b1}\,\dot{\theta}_5 - R_{b2}\,\dot{\theta}_{10} - \dot{y}_5\,cos\varphi + \dot{y}_{10}\,cos\varphi + \dot{x}_5\,sin\varphi - \dot{x}_{10}\,sin\varphi + \dot{e}_t\right) \tag{5.4}$$

where R_{b1} and R_{b2} are the basic radius of pinion and gear, and φ is the contact pressure angle.

The contact force is applied at nodes 5 and 10, considering the radius of the two gears and the contact angle:

$$\mathbf{f} = [T_{mot}, 0, 0, -R_{b1}\left(F_k + F_c\right), 0, 0, R_{b2}\left(F_k + F_c\right), 0, 0, 0, cos\varphi\left(F_k + F_c\right), 0, 0,$$

$$-cos\varphi\left(F_k + F_c\right), 0, 0, -sin\varphi\left(F_k + F_c\right), 0, 0, sin\varphi\left(F_k + F_c\right), 0]^T. \tag{5.5}$$

The whole torsional and translational motion equations are given as follows:

Torsional DOFs:

$$I_1\,\ddot{\theta}_1 = T_{mot} - k_{t12}\left(\theta_1 - \theta_2\right) - C_{t12}\left(\dot{\theta}_1 - \dot{\theta}_2\right) \tag{5.6-a}$$

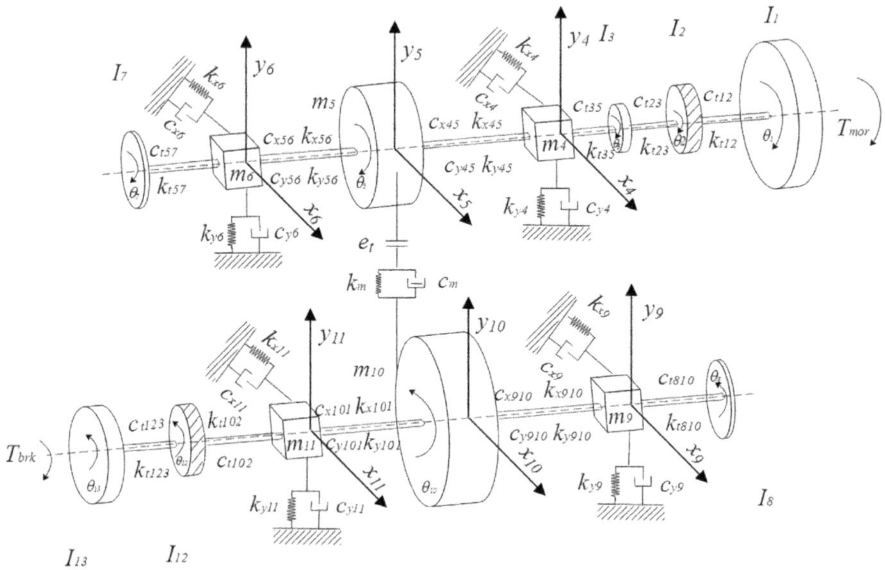

FIGURE 5.2 21 DOFs lumped dynamic model of UNSW test rig. Made by the author [14].

$$I_2\ddot{\theta}_2 = k_{t12}\left(\theta_1 - \theta_2\right) + C_{t12}\left(\dot{\theta}_1 - \dot{\theta}_2\right) - k_{t23}\left(\theta_2 - \theta_3\right) - C_{t23}\left(\dot{\theta}_2 - \dot{\theta}_3\right) \quad \text{(5.6-b)}$$

$$I_3\ddot{\theta}_3 = k_{t23}\left(\theta_2 - \theta_3\right) + C_{t23}\left(\dot{\theta}_2 - \dot{\theta}_3\right) - k_{t35}\left(\theta_3 - \theta_5\right) - C_{t35}\left(\dot{\theta}_3 - \dot{\theta}_5\right) \quad \text{(5.6-c)}$$

$$I_5\ddot{\theta}_5 = k_{t35}\left(\theta_3 - \theta_5\right) + C_{t35}\left(\dot{\theta}_3 - \dot{\theta}_5\right) + k_{t57}\left(\theta_7 - \theta_5\right) + C_{t57}\left(\dot{\theta}_7 - \dot{\theta}_5\right)$$
$$-R_{b1}\left(F_k + F_c\right) \quad \text{(5.6-d)}$$

$$I_7\ddot{\theta}_7 = k_{t57}\left(\theta_5 - \theta_7\right) + C_{t57}\left(\dot{\theta}_5 - \dot{\theta}_7\right) \quad \text{(5.6-e)}$$

$$I_8\ddot{\theta}_8 = k_{t810}\left(\theta_{10} - \theta_8\right) + C_{t810}\left(\dot{\theta}_{10} - \dot{\theta}_8\right) \quad \text{(5.6-f)}$$

$$I_{10}\ddot{\theta}_{10} = R_{b2}\left(F_k + F_c\right) - k_{t102}\left(\theta_{10} - \theta_{12}\right) - C_{t102}\left(\dot{\theta}_{10} - \dot{\theta}_{12}\right) + k_{t810}\left(\theta_8 - \theta_{10}\right)$$

$$+C_{t810}\left(\dot{\theta}_8 - \dot{\theta}_{10}\right) \quad (5.6\text{-g})$$

$$I_{12}\ddot{\theta}_{12} = k_{t102}\left(\theta_{10} - \theta_{12}\right) + C_{t102}\left(\dot{\theta}_{10} - \dot{\theta}_{12}\right) - k_{t123}\left(\theta_{12} - \theta_{13}\right) - C_{t123}\left(\dot{\theta}_{12} - \dot{\theta}_{13}\right)$$

(5.6-h)

$$I_{13}\ddot{\theta}_{13} = k_{t123}\left(\theta_{12} - \theta_{13}\right) + C_{t123}\left(\dot{\theta}_{12} - \dot{\theta}_{13}\right) - T_{brk} \quad (5.6\text{-i})$$

Translational DOFs:

$$m_4\ddot{y}_4 + k_{y4}y_4 + C_{y4}\dot{y}_4 + k_{y45}\left(y_4 - y_5\right) + C_{y45}\left(\dot{y}_4 - \dot{y}_5\right) = 0 \quad (5.7\text{-a})$$

$$m_5\ddot{y}_5 + k_{y56}\left(y_5 - y_6\right) + C_{y56}\left(\dot{y}_5 - \dot{y}_6\right) + k_{y45}\left(y_5 - y_4\right) + C_{y45}\left(\dot{y}_5 - \dot{y}_4\right) \quad (5.7\text{-b})$$
$$= (F_k + F_c)cos\varphi$$

$$m_6\ddot{y}_6 + k_{y6}y_6 + C_{y6}\dot{y}_6 - k_{y56}\left(y_5 - y_6\right) - C_{y56}\left(\dot{y}_5 - \dot{y}_6\right) = 0 \quad (5.7\text{-c})$$

$$m_9\ddot{y}_9 + k_{y9}y_9 + C_{y9}\dot{y}_9 + k_{y910}\left(y_9 - y_{10}\right) + C_{y910}\left(\dot{y}_9 - \dot{y}_{10}\right) = 0 \quad (5.7\text{-d})$$

$$m_{10}\ddot{y}_{10} + k_{y910}\left(y_{10} - y_9\right) + C_{y910}\left(\dot{y}_{10} - \dot{y}_9\right) + k_{y101}\left(y_{10} - y_{11}\right) + C_{y101}\left(\dot{y}_{10} - \dot{y}_{11}\right)$$
$$= \left(-F_k - F_c\right)cos\varphi \quad (5.7\text{-e})$$

$$m_{11}\ddot{y}_{11} + k_{y11}y_{11} + C_{y11}\dot{y}_{11} - k_{y101}\left(y_{10} - y_{11}\right) - C_{y101}\left(\dot{y}_{10} - \dot{y}_{11}\right) = 0 \quad (5.7\text{-f})$$

$$m_4 \ddot{x}_4 + k_{x4} x_4 + C_{x4} \dot{x}_4 + k_{x45}(x_4 - x_5) + C_{x45}\left(\dot{x}_4 - \dot{x}_5\right) = 0 \tag{5.7-g}$$

$$m_5 \ddot{x}_5 + k_{x56}(x_5 - x_6) + C_{x56}\left(\dot{x}_5 - \dot{x}_6\right) + k_{x45}(x_5 - x_4) + C_{x45}\left(\dot{x}_5 - \dot{x}_4\right) = (-F_k - F_c) sin\varphi \tag{5.7-h}$$

$$m_6 \ddot{x}_6 + k_{x6} x_6 + C_{x6} \dot{x}_6 - k_{x56}(x_5 - x_6) - C_{x56}\left(\dot{x}_5 - \dot{x}_6\right) = 0 \tag{5.7-i}$$

$$m_9 \ddot{x}_9 + k_{x9} x_9 + C_{x9} \dot{x}_9 + k_{x910}(x_9 - x_{10}) + C_{x910}\left(\dot{x}_9 - \dot{x}_{10}\right) = 0 \tag{5.7-j}$$

$$m_{10} \ddot{x}_{10} + k_{x910}(x_{10} - x_9) + C_{x910}\left(\dot{x}_{10} - \dot{x}_9\right) + k_{x101}(x_{10} - x_{11}) + C_{x101}\left(\dot{x}_{10} - \dot{x}_{11}\right) = \left(F_k + F_c\right) sin\varphi \tag{5.7-k}$$

$$m_{11} \ddot{x}_{11} + k_{x11} x_{11} + C_{x11} \dot{x}_{11} - k_{x101}(x_{10} - x_{11}) - C_{x101}\left(\dot{x}_{10} - \dot{x}_{11}\right) = 0 \tag{5.7-l}$$

5.2.1 Meshing Stiffness and Damping Coefficient of Gear System

In the developed dynamic model, the internal excitation of the gear systems is from the gear meshing contact [15–17]; therefore, properly modelling the contact properties of mating gear teeth is essential for a better understanding of the coupling effects between gear wear and gear dynamic characteristics, and then provide valuable information for wear monitoring and prediction.

Three major parameters are characterising the gear contact, that is, meshing stiffness k_m, damping coefficient c_m, and geometric transmission error e_t. In this book, the meshing stiffness k_m is considered as a function of the angular rotation of the pinion θ_5. The shape of the dependency $k_m\left(\theta_5\right)$ is derived based on the potential energy, which considers the strain energy and Hertzian contact [18–20]. The strain energy consists of bending, shear, and axial compressive potential energies stored in the meshing teeth, which are expressed below [21–23]:

$$U_b = \frac{F^2}{2k_b} \tag{5.8}$$

$$U_s = \frac{F^2}{2k_s} \tag{5.9}$$

$$U_a = \frac{F^2}{2k_a} \tag{5.10}$$

TABLE 5.2
Mass and Inertia of the Dynamic Model of Spur Gearbox Systems

Inertia (kgm^2)		**Mass (kg)**	
I_1	Inertia of motor, inertia of motor shaft and $1/2$inertia of coupling 1	m_4	Mass of pedestal and bearing
I_2	$1/2$inertia of coupling $1, 1/2$ inertia of coupling 2 and inertia of torque meter shaft	m_5	Mass of pinion
I_3	$1/2$inertia of coupling 2 and inertia of input shaft to bearing (not including bearing)	m_6	Mass of pedestal and bearing
I_5	Inertia of pinion, inertia of input shaft (whole shaft section through casing)	m_9	Mass of pedestal and bearing
I_7	Inertia of slip ring (rotor part), adapter, coupling 3 and shaft section outside casing (free end)	m_{10}	Mass of gear
I_8	Inertia of slip ring (rotor part), adapter, coupling 5 and shaft section outside casing (free end)	m_{11}	Mass of pedestal and bearing
I_{10}	Inertia of gear, inertia of whole output shaft section through casing		
I_{12}	$1/2$inertia of coupling 4 and output shaft section to bearing (not including bearing)		
I_{13}	Inertia of brake, inertia of brake shaft and $1/2$inertia of coupling 4		

where F is gear contact force, k_b, k_s and k_a are bending stiffness, shear stiffness, and axial compressive stiffness.

The stiffness of the Hertzian contact between tooth pairs can be approximated by [24]:

$$k_h = \frac{\pi E W}{4\left(1-\nu^2\right)}. \tag{5.11}$$

TABLE 5.3
Stiffness and Damping of the Dynamic Model of Spur Gearbox Systems

Stiffness and damping			
k_{t12}, C_{t12}	Torsional stiffness and damping of coupling 1	k_{y56}, C_{y56}	Vertical stiffness and damping of shaft
k_{t23}, C_{t23}	Torsional stiffness and damping of coupling 2	k_{y101}, C_{y101}	Vertical stiffness and damping of shaft
k_{t35}, C_{t35}	Torsional stiffness and damping of shaft (from coupling 2 to pinion)	k_{y910}, C_{y910}	Vertical stiffness and damping of shaft
k_{t57}, C_{t57}	Torsional stiffness and damping of shaft (from pinion to slip ring)	k_{x45}, C_{x45}	Horizontal stiffness and damping of shaft
k_{t810}, C_{t810}	Torsional stiffness and damping of shaft (from gear to slip ring)	k_{x56}, C_{x56}	Horizontal stiffness and damping of shaft
k_{t102}, C_{t102}	Torsional stiffness and damping of shaft (from gear to coupling 4)	k_{x101}, C_{x101}	Horizontal stiffness and damping of shaft
k_{t123}, C_{t123}	Torsional stiffness and damping of coupling 4	k_{x910}, C_{x45}	Horizontal stiffness and damping of shaft
k_{y45}, C_{y45}	Vertical stiffness and damping of shaft	$k_{y4}, k_{y6}, k_{y9}, k_{y11}$	Vertical stiffness of pedestal/bearing

The Hertzian stiffness is dependent on the width of contact between two teeth widths W and the material properties: Poisson's ratio and elastic modulus, v and E, respectively. With Hertzian, bending, shear, and axial compressive stiffness, the gear mesh stiffness for one tooth pair can be obtained by using [21, 25–27]:

$$\frac{1}{k_m} = \frac{1}{k_h} + \sum_{j=1}^{2} \left[\frac{1}{k_{b,j}} + \frac{1}{k_{s,j}} + \frac{1}{k_{a,j}} \right] \tag{5.12}$$

where $j = 1,2$represents the pinion and gear, respectively. The mesh stiffness of one tooth pair is demonstrated in Figure 5.3.

It is assumed that the gear system has a constant damping ratio, and the meshing damping coefficient is proportional to the gear meshing stiffness [28]:

$$c_m = \mu k_m \tag{5.13}$$

where $\mu(s)$is the scale constant.

To determine the value of μ, k_{ave} and c_{ave} are defined as the mean meshing stiffness and damping coefficient in one mesh period, and the damping ratio can be calculated through:

$$\zeta = \frac{c_{ave}}{2\sqrt{k_{ave} m}} \tag{5.14}$$

where m represents the effective mass of the pinion and gear, which can be derived as:

$$m = \frac{m_5 m_{10}}{m_5 + m_{10}}. \tag{5.15}$$

From Eq. (5.14) and Eq. (5.15), the average meshing damping coefficient c_{ave} is:

$$c_{ave} = 2\zeta\sqrt{k_{ave}\frac{m_5 m_{10}}{m_5 + m_{10}}}. \tag{5.16}$$

Therefore, based on Eqs. (5.13–5.16), the damping ratio μ together with the meshing damping coefficient c_m can be calculated.

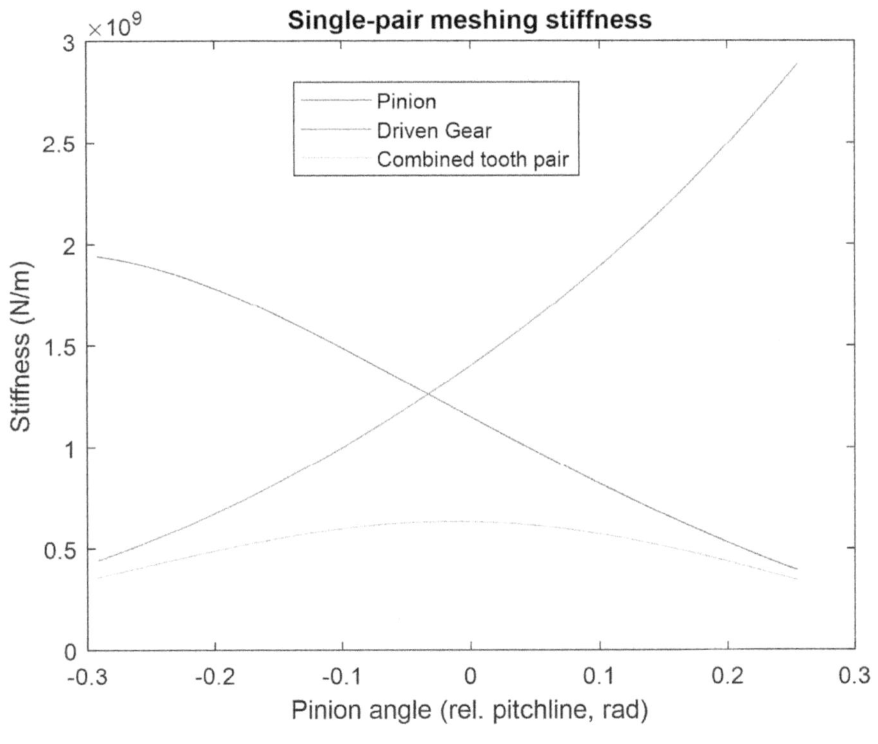

FIGURE 5.3 Single tooth pair meshing stiffness. Made by the author.

5.2.2 Geometric Transmission Error

The GTE, e_t (geometric deviation from perfect involute), makes *kinematic* (inertia and loading independent) contribution to the transmission error of the gear system [29]. It can be used to represent gear surface wear, but could also include initial profile errors which can be estimated based on manufacturing quality or measured.
The shape and severity of GTE impact the contact mechanism of gear pairs significantly. Figure 5.4 demonstrates the effects of GTE on gear dynamic contact force F_k. The deflection $\Delta\ell$ between the mating gear teeth is shown in Figure 5.4(b), it consists of the stiffness-related deformation and the geometric working surface deviations from the ideal gear tooth profile, which is GTE. Note that in the gear wear case, the meshing stiffness change can be neglected as it is significantly less important than the transmission error effect [11]. Therefore, the assessment of gear wear-induced GTE is vital for analysing the coupling effects between gear wear and gear dynamic, and the further wear prediction. The evaluation of wear-induced GTE will be introduced in Chapter 6.

5.2.3 Dynamic Simulation Process

In this book, the dynamic model is established using Simulink@ environment. To solve this developed gear system, the state space theory is used to represent the gear dynamic system in a useful mathematical way [30–32], as shown in Figure 5.5. Linear Time-Invariant (LTI) component is a linear representation of a dynamic gear system in either discrete or continuous time. With the LTI component, the eigenfrequencies can be easily obtained for validating the dynamic model. Also, the time-varying

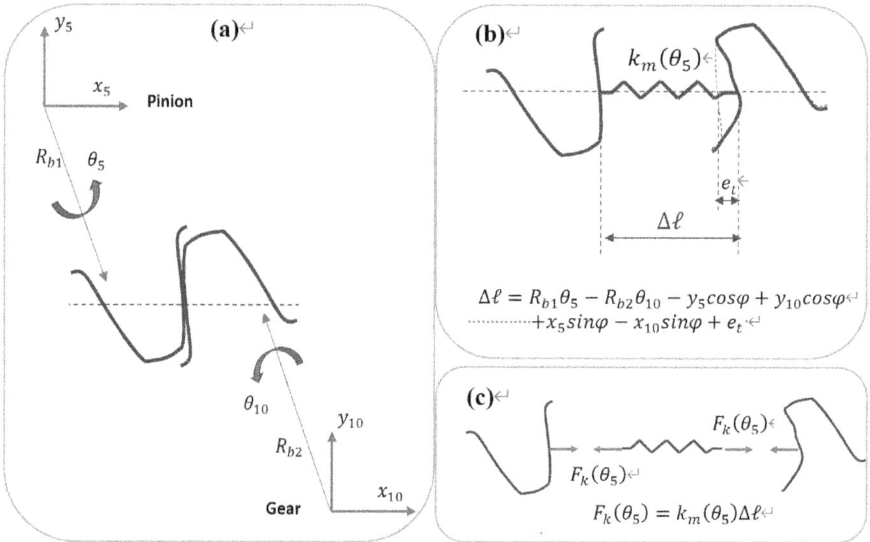

FIGURE 5.4 Effects of GTE on gear contact. Made by the author.

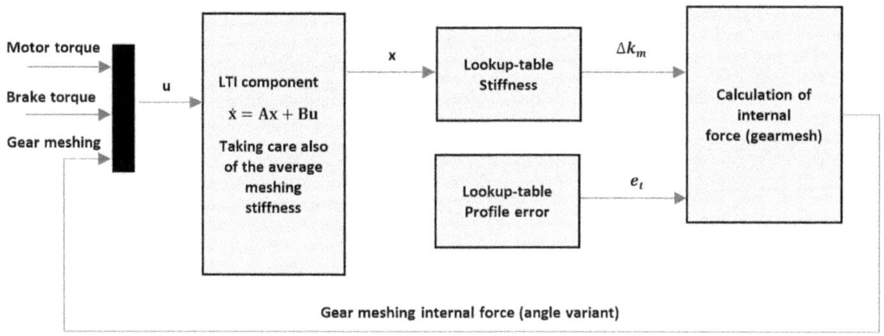

FIGURE 5.5 Gear state-space dynamic model. Made by the author.

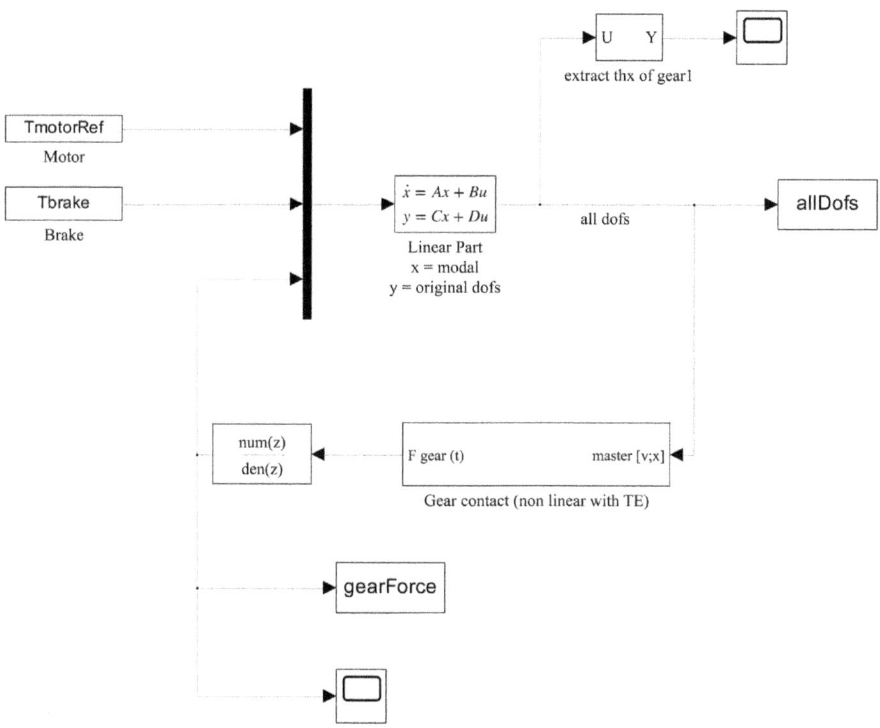

FIGURE 5.6 Dynamic model in Simulink®: Overall structure. Made by the author.

variables, such as meshing stiffness, damping coefficient, and geometric transmission error, are taken into consideration, using the lookup tables with cubic spline interpolation, so that the dynamic characteristics of the gear system can be realised and simulated. To clearly show the differences between this book and previously published research [11–13,33], more details of the dynamic model in the Simulink® simulation environment are demonstrated in Figures 5.6, 5.7, and 5.8.

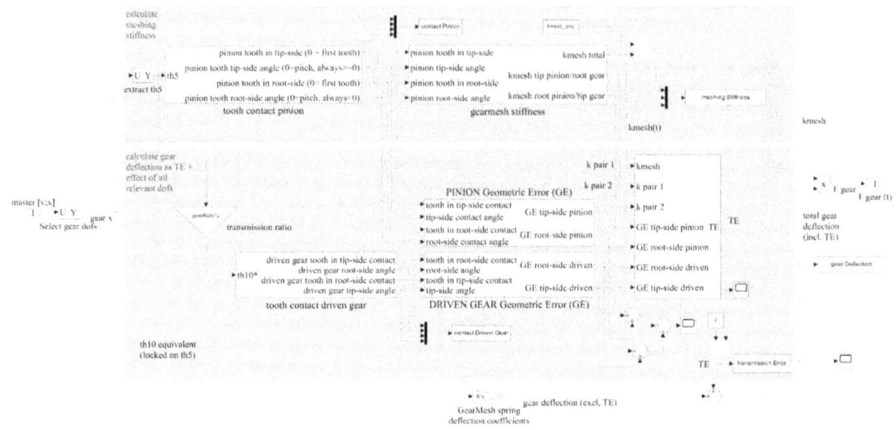

FIGURE 5.7 Dynamic model in Simulink®: Gear contact part. Made by the author.

ODE 45 is a function that implements a Runge–Kutta method with a variable time step and high computation efficiency; therefore, in this book, ODE45 would be the first choice for solving the dynamic motion equations [12]. In some specific cases, for example, the problem/system is suspected to be stiff [34], ODE15s is the alternative in this book to obtain numerical solutions of the developed dynamic gear transmission system.

5.3 MODEL VALIDATION AND CALIBRATION

Before using the dynamic model for further analysis, the dynamic model should be validated and calibrated to guarantee that the responses from the dynamic model are reliable. In this book, a series of initial tests were performed for the validation and calibration of the dynamic model, including a speed ramp test, impact tests, and constant speed tests.

To obtain the natural frequencies/modes of the spur gearbox test rig for comparison with the developed dynamic gear model, a speed ramp test was arranged. This method involves collecting vibration signals over a period when the gearbox is ramping up toward full speed, as shown in Figure 5.9. The vibration of the shaft acts as a forcing function, supplying energy to the gear system and exciting resonances as the shaft vibration passes through the critical speed.

Power spectral density (PSD) analysis is a traditional frequency-domain analysis tool to identify the structural modes/natural frequencies of mechanical systems. The reason is that resonance is the amplification of a signal when its frequency is close to the natural frequency of a system. With the help of PSD analysis, the resonances in the spectrum are amplified, and other components, such as background noise, are reduced, thereby allowing the system modes to be more easily detected. The PSD analysis of the spur gearbox is shown in Figure 5.10.

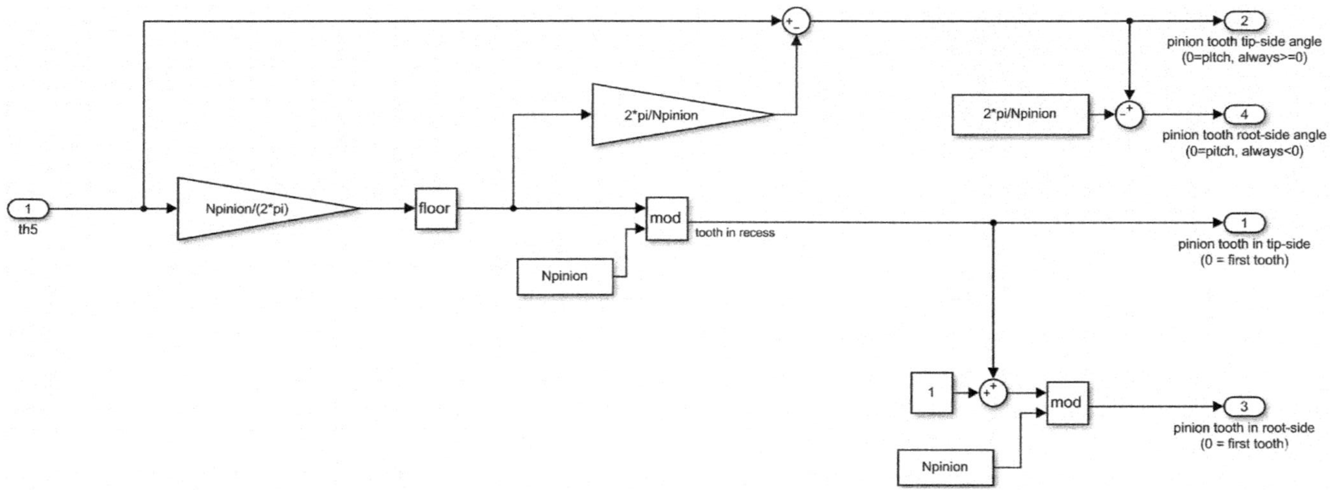

FIGURE 5.8 Dynamic model in Simulink®: Single tooth pair contact. Made by the author.

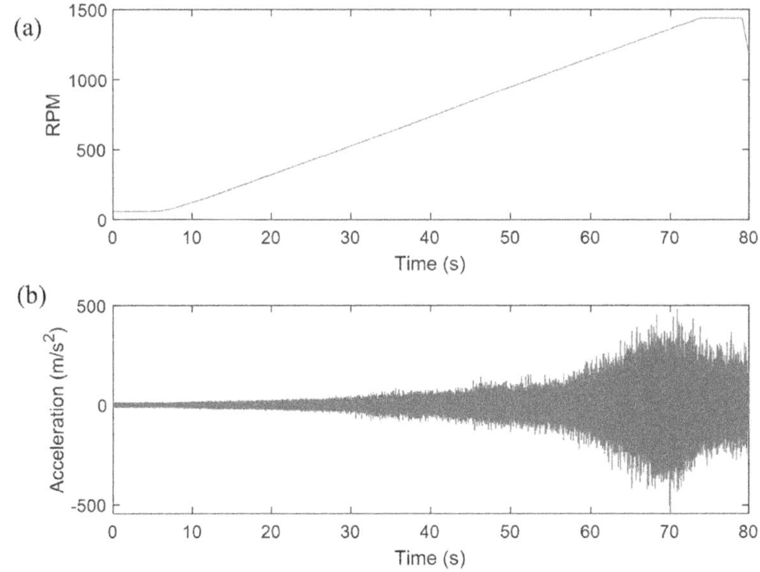

FIGURE 5.9 Measured vibration during the spur gearbox ramp test.

From Figure 5.10, the modes/natural frequencies of the whole UNSW spur gear system can be identified, which have been summarised in Table 5.4. These natural frequencies can be used as references to help check the validity and reliability of the developed dynamic model. It should be noted that in the developed dynamic model, the foundation of the spur gearbox is not simulated and included; therefore, to make the natural frequency from ramp tests comparable with the developed dynamic model of the gear system, the foundation of the gearbox's natural frequencies should be excluded from the natural frequencies acquired from ramp tests. With this regard, an impact test was applied to help reveal the natural frequencies of the gearbox's foundation so that it can be subtracted from the whole UNSW spur gear system. Note that a slight load is applied to ensure the gear pairs are in contact, when conducting the hammer test on the foundation of the gearbox.

The quadrature picking method was applied here to determine the natural frequencies/modes and mode shapes of the gearbox's foundation [35], as demonstrated in Figure 5.11. The theory of the quadrature picking method is explained as follows. The Frequency Response Function (FRF) appears to become purely imaginary at the modal frequency. Its amplitude is proportional to the modal displacement, and its sign is positive if the displacement is in phase with the excitation. The mode shapes can be determined if a response is fixed, or set an excitation degree of freedom as a reference, and then make a set of measurements. The imaginary parts of the measured FRFs can be 'picked' at the modal frequencies at which they represent the modal displacement for that specific degree of freedom [35]. This method is based on the assumption that the coupling between the modes is light. In practice, mechanical structures are often very lightly damped (<1%). This implies that the modes are lightly coupled.

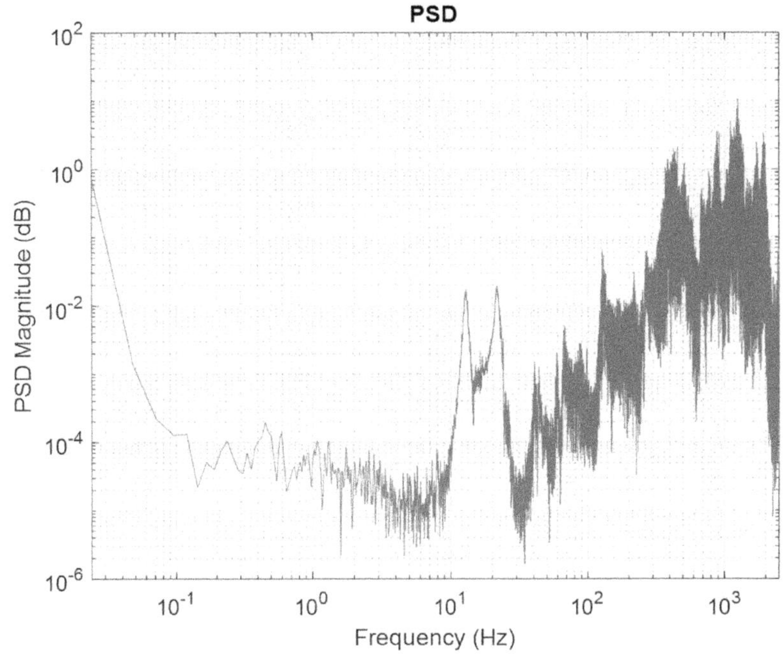

FIGURE 5.10 PSD analysis of the vibrations of the spur gear system.

TABLE 5.4
Identified Natural Frequencies from the PSD Analysis of Experimental Measurements

Natural frequencies (Hz)						
13	22	41	59	68	87	128
218	270	430	512	6883	767	880
1103	1167	1248	1367	1471	1714	1927

As for the foundation of the gear system, an accelerometer (B&K 4396) is fixed on the root of the gearbox casing to collect the responses excited by varying inputs generated by impact hammers, as shown in Figure 5.12. The modes of the foundation of the gearbox and its mode shapes can be calculated and extracted through Eq. (5.17):

$$H(\omega) = \frac{X(\omega)}{F(\omega)} \tag{5.17}$$

where $H(\omega)$ is the FRF, $X(\omega)$ is the output of the system and $F(\omega)$ is the input of the system.

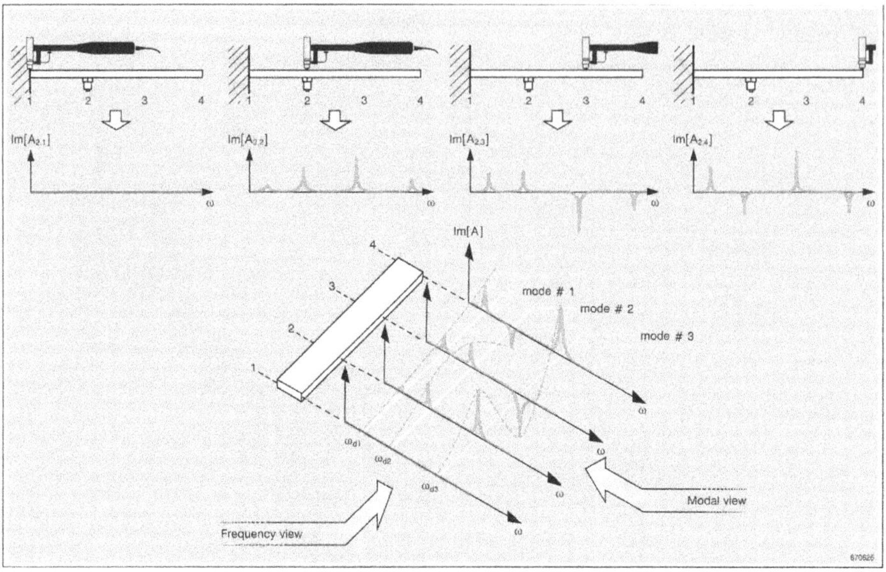

FIGURE 5.11 Demonstration of the quadrature picking method [35]. Made by Brüel.

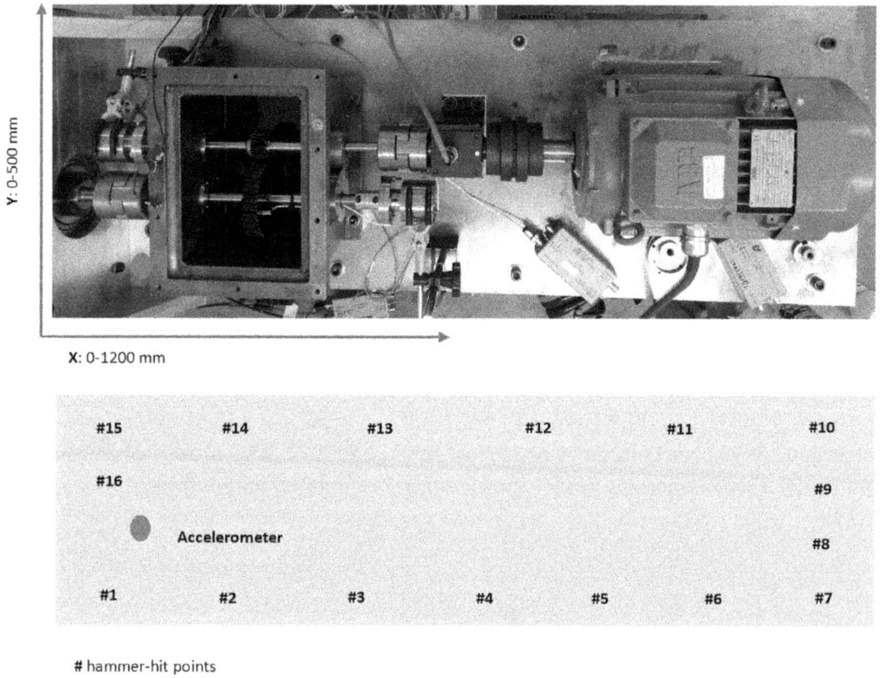

FIGURE 5.12 Hammer test on gearbox's foundation. Photograph by the author.

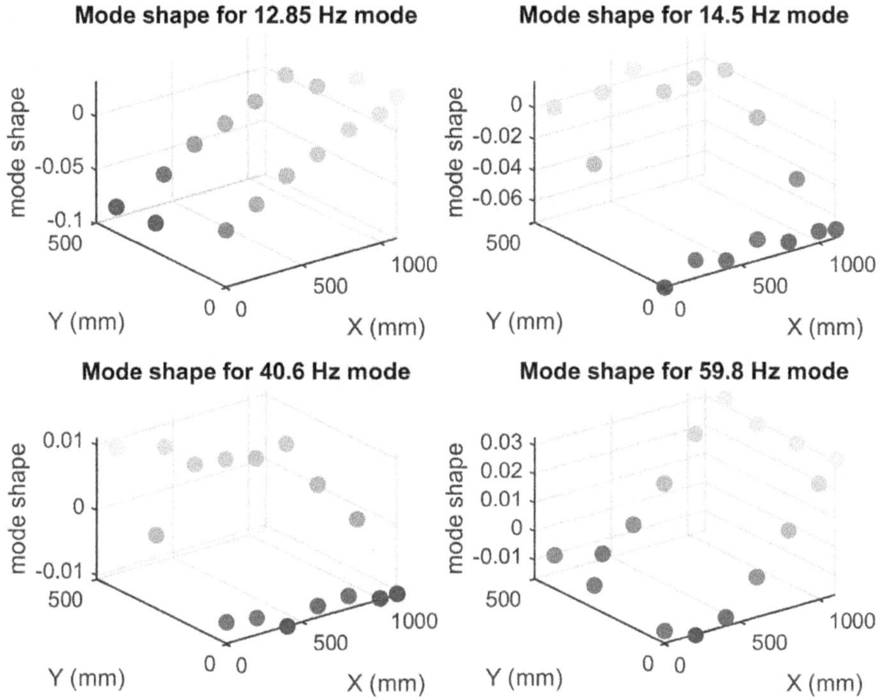

FIGURE 5.13 The first four mode shapes of the gearbox's foundation. Made by the author.

Figure 5.13 gives the first four mode shapes for demonstration purposes. Based on Eq. (5.17), the FRF of the gearbox's foundation is shown in Figure 5.14, including the magnitude, phase, and coherence. At the modes of the gear system, there would be a peak in the magnitude spectrum, and the corresponding phase would shift 180 degrees. Coherence is a function versus frequency that indicates how much of the output is due to the input in the FRF. It can be an indicator of the quality of the FRF, which evaluates the consistency of the FRF from measurement to repeat of the same measurement: (a) coherence's value is 1 at a particular frequency indicating that the FRF amplitude and phase are very repeatable from measurement to measurement; (b) while coherence's value is 0 indicating that opposite – the measurements are not repeatable, which is a possible 'warning flag' that there is an error in the measurement setup.

From the magnitude, phase, and coherence spectrums in Figure 5.14, all the modes of the gearbox's foundation can be identified. Subtracting the obtained modes/natural frequencies from the whole gear system's modes (from the speed ramp test), the modes/natural frequencies of the gear system without foundation can be achieved, which are comparable with the developed dynamic model.

The speed ramp and impact tests were used to provide an initial adjustment of the most uncertain model parameters (e.g. stiffness of joints and bearings) with the ultimate aim of obtaining a good match between simulated and experimental natural frequencies in the frequency range of interest (0–2 kHz). The comparison result is summarised

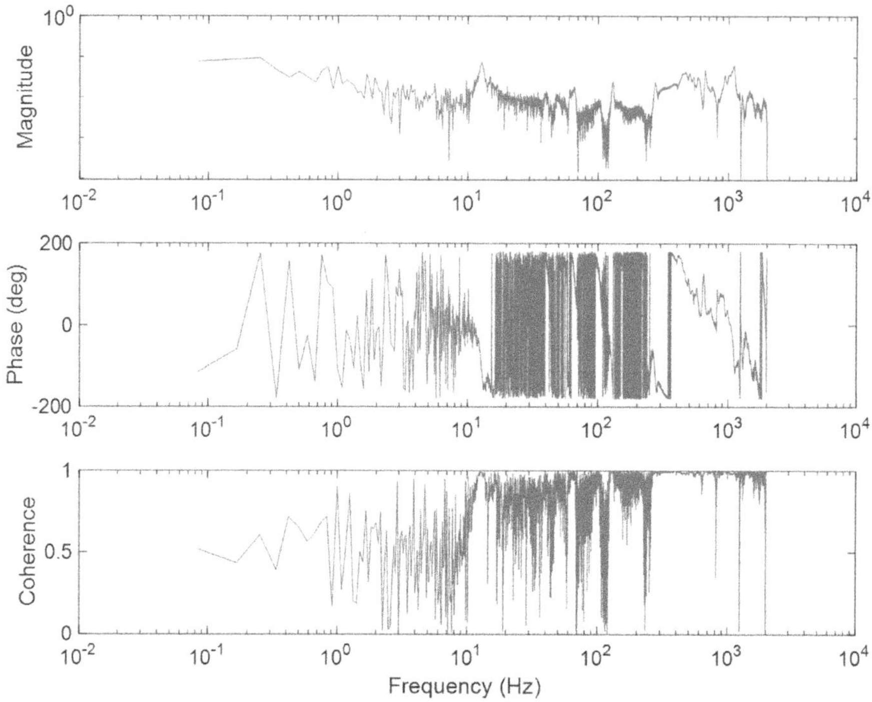

FIGURE 5.14 Frequency Response Function (FRF) of the gearbox's foundation. Made by the author.

in Table 5.5, and the scaled meshing stiffness is shown in Figure 5.15. Note that the natural frequencies presenting formats (such as accuracy) are set to be the same as those of the researchers in reference [36]. The reason why scaling the meshing stiffness is explained as follows. In gear transmission systems, various gears are used for different purposes. For example, some gears are surface hardened, while some are not treated with the hardening process. Soft vs hardened gears result in different magnitudes of the meshing stiffness when they have identical gear profiles. To guarantee the responses from the dynamic model are close to the actual measurements from the specific test rig, model calibration is thus necessary. It involves 'scaling' the meshing stiffness of the model so that it is close enough to the gears used in the experiments.

Table 5.5 shows that the first eight natural frequencies agree between the experiments and the dynamic model within eight percent, which was deemed sufficient to approximate the system response.

Two constant speed tests were then executed to fine-tune a scale factor for the meshing stiffness function k_m and the damping coefficient c_m:

1. 2 Hz rotational speed and 10 Nm motor torque.
2. 2 Hz rotational speed and 20 Nm motor torque.

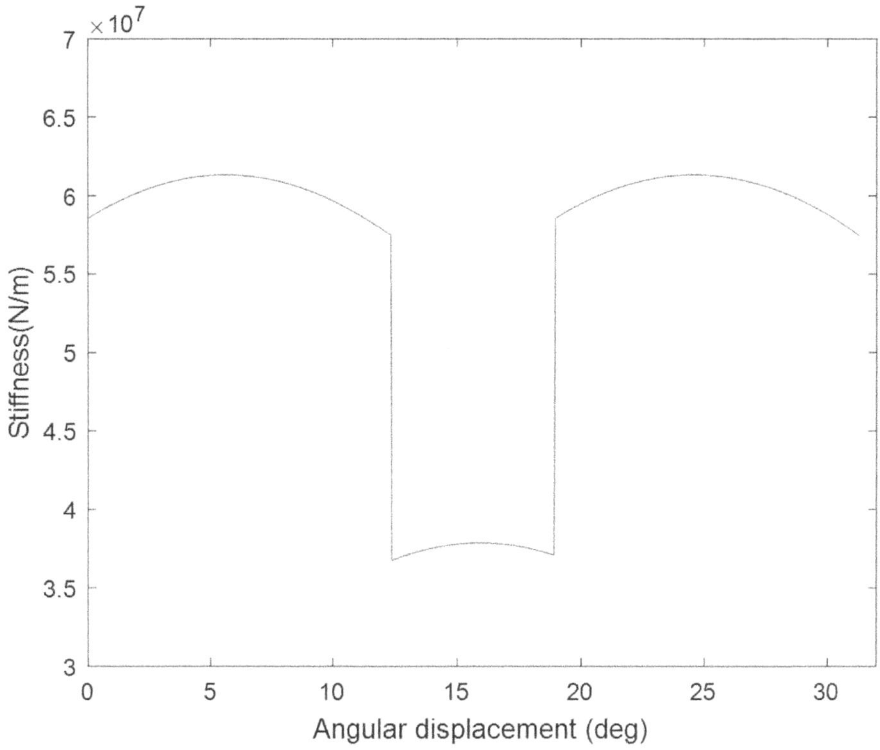

FIGURE 5.15 Meshing stiffness curve. Made by the author [37].

TABLE 5.5
Natural Frequency Comparison Results of Dynamic Model and Experimental Data

Experiments (Hz)	Dynamic model (Hz)	Difference (%)
22	23	5.1
87	81	-6.7
218	201	-7.8
430	427	-0.8
512	529	3.5
767	740	3.5
1248	1268	1.6
1927	1885	2.2

The encoder signals on the input and output shaft were used in these tests to calculate the static transmission error (STE) of the gearbox system using the phase demodulation method [38] assuming negligible dynamic effects at this low speed. This STE was then used as an input to the dynamic model to simulate a vibration response. After low-pass filtering in the band of interest (0–2 kHz), the RMS of the simulated vibration signal $y_6^{(SIM)}(t)$ was then compared with the experimental results, and the meshing stiffness and damping were manually adjusted until good agreement between the results was obtained, as shown in Figure 5.16. To examine the rationality of the responses from the developed dynamic model, similar to reference [12], the first five gear tooth meshes are expanded and plotted in Figure 5.17. From the comparison between the five gear mesh teeth signals of the experiment and simulation, it can be seen that the characteristics of the gear system are well presented, including the engagement of teeth contact, loading and unloading of teeth, the gear mesh period, and the amplitude of vibration. Note that the wear on each gear tooth is theoretically the same, even though the tooth-to-tooth differences do exist in practice due to manufacturing or mounting error (as shown in Figure 5.16(a) and Figure 5.17(a)), resulting in that modulation phenomenon and sidebands exist in the measured signal. Since uniform wear is simulated in the developed dynamic model, and also because the wear process is the same on each gear tooth, no sidebands around the gear mesh (caused by the modulations behaviours) exist in the simulated signal.

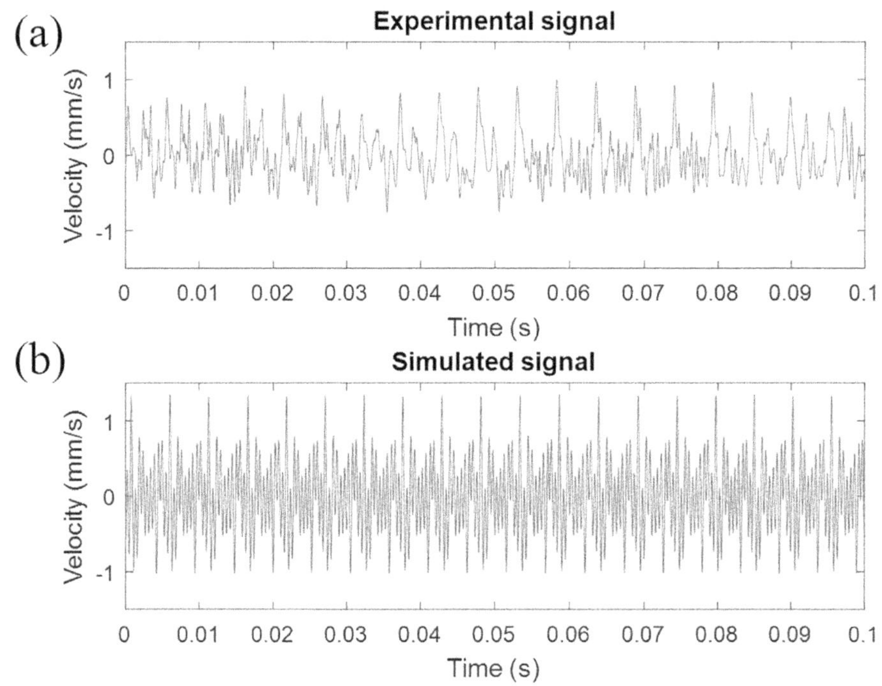

FIGURE 5.16 Vibration signal: Experiment and simulation. Made by the author [37].

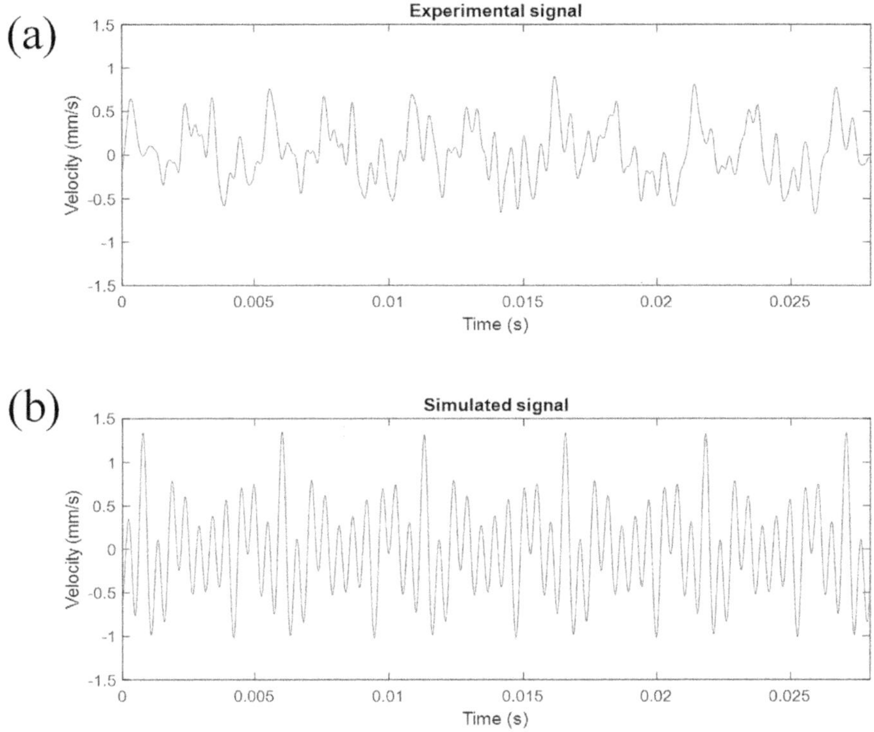

FIGURE 5.17 Five-gearmesh teeth: Experiment and simulation. Made by the author.

The low-pass filtering was necessary to remove high-frequency effects which arise in the simulation from the unrealistic steps in the stiffness function, smoothed in practice by a gradual engagement/disengagement of the teeth. Indeed, even in this limited band, the stronger weighting of the higher gear mesh harmonics is apparent in the simulated response in Figure 5.16(b). Nonetheless, the overall magnitudes of the responses were deemed to match the measurements sufficiently for the purposes of the present work.

The dynamic contact force from the developed model is shown in Figure 5.18, and more spikes and fluctuations are observed in the contact force compared with the results shown in reference [10]. These spikes and fluctuations are caused by the resonances of the gear system. In reference [10], a torsional model (with a single DOF) was established to generate the dynamic contact force and the subsequent worn tooth profile. As for the torsional model in reference [10], there is only one mode existing in the simulated gear system; therefore, its FRF is simple and only has one peak, resulting in the contact force being in a ‘smooth’ pattern. However, in practice, the gear system is usually much more complex in terms of potential changes in the dynamic responses under an operating condition, and the single DOF torsional model is not representative anymore. Therefore, the authors in reference [39] included the translational effects coming from the shaft bending and bearing radial deflections

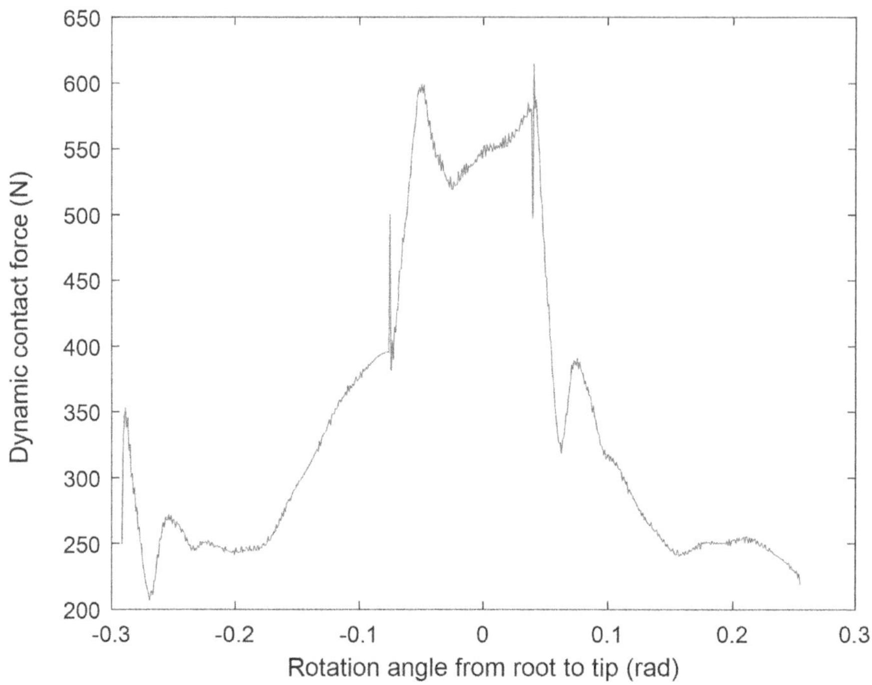

FIGURE 5.18 Dynamic contact force (from gear root to tip). Made by the author.

into the dynamic model, and significantly different characteristics of contact force and the subsequent worn tooth profile were observed, compared with the results shown in reference [10], and more spikes and fluctuations start to appear. Also, the dynamic contact force simulated in reference [40] exhibited lots of spikes and fluctuation behaviours due to more degrees of freedom are included in the developed dynamic model.

This calibrated dynamic model was used both to generate dynamic contact forces (for specified GTEs) for input into the tribological wear model and to simulate responses for comparison with vibration measurements to enable updating of the wear model.

5.4 SUMMARY

This chapter presents the development of a 21-DOFs dynamic model based on the UNSW spur gearbox test rig, including gear meshing stiffness, geometric transmission error evolution, and the solver to acquire numerical solutions. The dynamic model developed in this chapter is a crucial part of the proposed vibration-based wear prediction schemes, which will be introduced in Chapter 6 and Chapter 7. Note that the developed dynamic model is different from previously published research

[11–13] since a new gearbox test rig is used in this book. Also, the applications of the developed dynamic model are different from those in previously published research. To guarantee that the dynamic responses of the developed dynamic model are close enough to the actual test rig so that it can provide useful information for further wear analysis, speed ramp test, impact test, and several constant speed tests are conducted to validate and calibrate the dynamic model. The natural frequency comparison between experiment and simulation shows that the developed dynamic model matches well with the real UNSW spur gearbox test rig. Also, the time waveforms of vibrations from experiments and the dynamic model match each other. It suggests that the developed dynamic model can provide reliable dynamic contact to the tribological (wear) models for further wear analysis. Moreover, the simulated vibration characteristics can be used to compare with actual measurements for updating the analysis of wear prediction, which will be introduced in Chapter 6 and Chapter 7.

However, it should be mentioned that the foundation of the gearbox has not been modelled in the developed dynamic model, which might affect the simulated vibrations, especially when it is operating in its resonance range. This issue will be addressed in future work by including the foundation.

REFERENCES

[1] S. Li, Finite element analyses for contact strength and bending strength of a pair of spur gears with machining errors, assembly errors and tooth modifications, *Mechanism and Machine Theory*, 42 (2007) 88–114. https://doi.org/10.1016/j.mechmachtheory.2006.01.009.

[2] J. Hedlund, A. Lehtovaara, Modeling of helical gear contact with tooth deflection, *Tribology International*, 40 (2007) 613–619. https://doi.org/10.1016/j.triboint.2005.11.004.

[3] V. Simon, FEM stress analysis in hypoid gears, *Mechanism and Machine Theory*, 35 (2000) 1197–1220. https://doi.org/10.1016/S0094-114X(99)00071-3.

[4] K. Hayashi, Load distribution on the contact line of helical gear teeth: Part 1 fundamental concept, *Bulletin of JSME*, 6 (1963) 336–343.

[5] J. Boerner, *Very efficient calculation of the load distribution on external gear sets – The method and applications of the program LVR*, Proc. of the 7th International Power Transmission and Gearing Conference, ASME San Diego, 1996, pp. 219–225.

[6] M. Pimsarn, K. Kazerounian, Efficient evaluation of spur gear tooth mesh load using pseudo-interference stiffness estimation method, *Mechanism and Machine Theory*, 37 (2002) 769–786. https://doi.org/10.1016/S0094-114X(02)00022-8.

[7] J.I. Pedrero, M. Pleguezuelos, M. Artés, J.A. Antona, Load distribution model along the line of contact for involute external gears, *Mechanism and Machine Theory*, 45 (2010) 780–794. https://doi.org/10.1016/j.mechmachtheory.2009.12.009.

[8] S. Li, Effect of addendum on contact strength, bending strength and basic performance parameters of a pair of spur gears, *Mechanism and Machine Theory*, 43 (2008) 1557–1584. https://doi.org/10.1016/j.mechmachtheory.2007.12.010.

[9] M. Arafa, M. Megahed, Evaluation of spur gear mesh compliance using the finite element method, *Proceedings of the Institution of Mechanical Engineers, Part C: Journal of Mechanical Engineering Science*, 213 (1999) 569–579.

[10] H. Ding, A. Kahraman, Interactions between nonlinear spur gear dynamics and surface wear, *Journal of Sound and Vibration*, 307 (2007) 662–679. https://doi.org/10.1016/j.jsv.2007.06.030.

[11] H. Endo, *Simulation of gear faults and its application to the development of differential diagnostic technique*, School of Mechanical and Manufacturing Engineering, The University of New South Wales, Australia, 2005.

[12] N. Sawalhi, R.B. Randall, Simulating gear and bearing interactions in the presence of faults: Part I. The combined gear bearing dynamic model and the simulation of localised bearing faults, *Mechanical Systems and Signal Processing*, 22 (2008) 1924–1951. https://doi.org/10.1016/j.ymssp.2007.12.001.

[13] L. Deshpande, N. Sawalhi, R. Randall, *Gearbox fault simulation using finite element model reduction technique*, Proceedings of Acoustics, Gold Coast, Australia, 2011, p. 111.

[14] K. Feng, W.A. Smith, R.B. Randall, H Wu, Z. Peng, Vibration-based monitoring and prediction of surface profile change and pitting density in a spur gear wear process, *Mechanical Systems and Signal Processing*, 165 (2022) 108319. https://doi.org/10.1016/j.ymssp.2021.108319.

[15] X. Liang, M.J. Zuo, Z. Feng, Dynamic modeling of gearbox faults: A review, *Mechanical Systems and Signal Processing*, 98 (2018) 852–876. https://doi.org/10.1016/j.ymssp.2017.05.024.

[16] Y. Huangfu, K. Chen, H. Ma, X. Li, H. Han, Z. Zhao, Meshing and dynamic characteristics analysis of spalled gear systems: A theoretical and experimental study, *Mechanical Systems and Signal Processing*, 139 (2020) 106640. https://doi.org/10.1016/j.ymssp.2020.106640.

[17] C.J. Li, H. Lee, Gear fatigue crack prognosis using embedded model, gear dynamic model and fracture mechanics, *Mechanical Systems and Signal Processing*, 19 (2005) 836–846. https://doi.org/10.1016/j.ymssp.2004.06.007.

[18] X. Liang, M.J. Zuo, M. Pandey, Analytically evaluating the influence of crack on the mesh stiffness of a planetary gear set, *Mechanism and Machine Theory*, 76 (2014) 20–38. https://doi.org/10.1016/j.mechmachtheory.2014.02.001.

[19] X. Liang, M.J. Zuo, T.H. Patel, Evaluating the time-varying mesh stiffness of a planetary gear set using the potential energy method, *Proceedings of the Institution of Mechanical Engineers, Part C: Journal of Mechanical Engineering Science*, 228 (2013) 535–547. https://doi.org/10.1177/0954406213486734.

[20] Z. Wan, H. Cao, Y. Zi, W. He, Y. Chen, Mesh stiffness calculation using an accumulated integral potential energy method and dynamic analysis of helical gears, *Mechanism and Machine Theory*, 92 (2015) 447–463. https://doi.org/10.1016/j.mechmachtheory.2015.06.011.

[21] D.C.H. Yang, J.Y. Lin, Hertzian Damping, Tooth friction and bending elasticity in gear impact dynamics, *Journal of Mechanisms, Transmissions, and Automation in Design*, 109 (1987) 189–196. https://doi.org/10.1115/1.3267437.

[22] K. Chen, Y. Huangfu, H. Ma, Z. Xu, X. Li, B. Wen, Calculation of mesh stiffness of spur gears considering complex foundation types and crack propagation paths, *Mechanical Systems and Signal Processing*, 130 (2019) 273–292. https://doi.org/10.1016/j.ymssp.2019.05.014.

[23] Y. Pandya, A. Parey, Experimental investigation of spur gear tooth mesh stiffness in the presence of crack using photoelasticity technique, *Engineering Failure Analysis*, 34 (2013) 488–500. https://doi.org/10.1016/j.engfailanal.2013.07.005.

[24] D.C.H. Yang, Z.S. Sun, A rotary model for spur gear dynamics, *Journal of Mechanisms, Transmissions, and Automation in Design*, 107 (1985) 529–535. https://doi.org/10.1115/1.3260759.

[25] X. Liang, H. Zhang, L. Liu, M.J. Zuo, The influence of tooth pitting on the mesh stiffness of a pair of external spur gears, *Mechanism and Machine Theory*, 106 (2016) 1–15. https://doi.org/10.1016/j.mechmachtheory.2016.08.005.

[26] L. Liu, X. Liang, M.J. Zuo, Vibration signal modeling of a planetary gear set with transmission path effect analysis, *Measurement*, 85 (2016) 20–31. https://doi.org/10.1016/j.measurement.2016.02.006.

[27] Y. Pandya, A. Parey, Failure path based modified gear mesh stiffness for spur gear pair with tooth root crack, *Engineering Failure Analysis*, 27 (2013) 286–296. https://doi.org/10.1016/j.engfailanal.2012.08.015.

[28] M. Amabili, A. Rivola, Dynamic analysis of spur gear pairs: Steady-state response and stability of the SDOF model with time-varying meshing damping, *Mechanical Systems and Signal Processing*, 11 (1997) 375–390. https://doi.org/10.1006/mssp.1996.0072.

[29] W.D. Mark, *Performance-based gear metrology*, Wiley Online Library, Chichester, West Sussex, United Kingdom, 2012.

[30] A.S. Kumar, T.S. Sankar, M.O.M. Osman, On dynamic tooth load and stability of a spur-gear system using the state-space approach, *Journal of Mechanisms, Transmissions, and Automation in Design*, 107 (1985) 54–60. https://doi.org/10.1115/1.3258695

[31] H. Nevzat Özgüven, D.R. Houser, Mathematical models used in gear dynamics – A review, *Journal of Sound and Vibration*, 121 (1988) 383–411. https://doi.org/10.1016/S0022-460X(88)80365-1.

[32] L. Yutao, T. Di, Dynamics modeling of planetary gear set considering meshing stiffness based on bond graph, *Procedia Engineering*, 24 (2011) 850–855. https://doi.org/10.1016/j.proeng.2011.11.2749.

[33] N. Sawalhi, *Diagnostics, prognostics and fault simulation for rolling element bearings*, School of Mechanical and Manufacturing Engineering, The University of New South Wales, Australia, 2007.

[34] X. Tian, *Dynamic simulation for system response of gearbox including localized gear faults*, Master of Science Thesis, Department of Mechanical Engineering, University of Alberta Edmonton, Canada, 2004.

[35] Brüel and Kjær. Modal analysis and simulation: Structural testing–Part II, Brüel and Kjær, Nærum, Denmark, 1988. Accessed on www.bksv.com/media/doc/br0507.pdf: Retrieved May 5, 2026.

[36] Y. Guo, T. Eritenel, T.M. Ericson, R.G. Parker, Vibro-acoustic propagation of gear dynamics in a gear-bearing-housing system, *Journal of Sound and Vibration*, 333 (2014) 5762–5785. https://doi.org/10.1016/j.jsv.2014.05.055.

[37] K. Feng, P. Borghesani, W.A. Smith, R.B. Randall, Z.Y. Chin, J. Ren, Z. Peng, Vibration-based updating of wear prediction for spur gears, *Wear*, 426–427 (2019) 1410–1415. https://doi.org/10.1016/j.wear.2019.01.017.

[38] P.J. Sweeney, R.B. Randall, Gear transmission error measurement using phase demodulation, *Proceedings of the Institution of Mechanical Engineers, Part C: Journal of Mechanical Engineering Science*, 210 (1996) 201–213. https://doi.org/10.1243/pime_proc_1996_210_190_02.

[39] X. Liu, Y. Yang, J. Zhang, Investigation on coupling effects between surface wear and dynamics in a spur gear system, *Tribology International*, 101 (2016) 383–394. https://doi.org/10.1016/j.triboint.2016.05.006.

[40] H. Dai, F. Chen, C. Xun, X. Long, Numerical calculation and experimental measurement for gear mesh force of planetary gear transmissions, *Mechanical Systems and Signal Processing*, 162 (2022) 108085. https://doi.org/10.1016/j.ymssp.2021.108085.

6 Wear-Induced Tooth Profile Change Monitoring and Prediction

6.1 INTRODUCTION

Gear wear is a gradual loss of material from contacting gear tooth surfaces due to the combined sliding and rolling motions under boundary or mixed lubrication conditions [1]. During a gear's service life, wear-induced changes in the tooth profile are common and can lead to a reduction in tooth thickness. Generally, the wear-induced tooth profile changes are unevenly distributed from the root to the tip of the gear teeth [2]. The reason is that the sliding velocities at the different contact locations of the gear tooth in relation to the pitch line are different, and the wear rate is a function of the sliding velocity as well as the pressure [3]. The wear-induced tooth profile alteration may lead to a sudden failure of the gearbox transmission system, which can result in unexpected economic loss and serious accidents. Therefore, for the effective management of the health of the transmission system, it is important to be able to monitor the changes in the gear profile and predict their propagation, which is the focus of this chapter.

As discussed in Chapter 2, most existing research has focused on studying the effects of surface wear processes on the dynamic characteristics of gear systems, such as transmission error and dynamic meshing force [4–8], as well as on investigating the effect of gear dynamics on surface wear [9–14]. In contrast, there are only a handful of studies on the prediction of spur gear wear-induced tooth profile change under quasi-static operating conditions [2, 15, 16]. However, dynamic response characteristics differ significantly from those under quasi-static conditions. Dynamic meshing forces are typically larger than their quasi-static counterparts, with significant differences in both magnitude and waveform [9].

Theoretically, dynamic response characteristics of a gear pair are sensitive to profile changes, such as geometric deviations from a perfect involute tooth surface and elastic deformation [17]. Gear surface wear is a material removal process that leads to a geometric deviation. Elastic deformation is determined by contact force and meshing stiffness. Based on this theory, the authors of reference [9] used a periodically time-varying meshing stiffness function and an external displacement excitation to represent the effects of dynamic response on the gear wear process. In that study, the authors employed a torsional model with a single-degree-of-freedom (DOF) and then combined it with a quasi-static wear model [15] to develop a dynamic wear

DOI: 10.1201/9781003686736-6

model. This model can be used to investigate the interactions between the surface wear and the spur gear system's dynamic characteristics. To demonstrate the effect of translational deflection on the wear process, a later study [11] introduced a 3-DOFs dynamic model to replace the torsional model in reference [9]. The study found that the translational deflection in the gear system impacts the gear wear progradation significantly, and it must be included when analysing gear wear progradation.

It should be noted that references [9, 11] primarily focused on the coupling effects between gear dynamics and the wear process through a set of simulations. Gear wear prediction under dynamic conditions, which offers significant benefits to various industries, was not addressed.

The combination of dynamic and wear models, proposed in a few variants in the literature [2, 15, 16], can theoretically predict the evolution of wear and the resulting tooth profile alterations, considering its interactions with the gearbox dynamics. However, this prediction is likely to diverge from the actual wear process, which is a complex process with multiple factors, especially considering that the parameters governing wear dynamics vary in time (e.g. with the contamination of the lubricant and the change in surface roughness). Increasing the complexity of the wear models to follow these complex trends is an option, but this could introduce additional parameters that are difficult to quantify for practical applications, thereby increasing uncertainty in the model predictions.

To accurately and efficiently predict gear tooth profile change in a wear process, a gear-wear prediction methodology is proposed in this book. The following sections will introduce the complete procedure of the proposed vibration-based updating of the wear prediction scheme and demonstrate the tooth profile change prediction results with accelerated run-to-failure gear wear tests under various lubrication conditions.

6.2 METHODOLOGY FOR MONITORING AND PREDICTING TOOTH PROFILE CHANGE FROM WEAR

6.2.1 The Proposed Vibration-Based Approach for Monitoring and Predicting Tooth Profile Change

This section briefly introduces the architecture of the proposed vibration-based prediction scheme for the tooth profile changes due to wear.

The overall approach is presented concisely in Figure 6.1. The modelling component of this methodology (on the left of Figure 6.1) consists of two interacting simulation models: A dynamic model and a wear model. Based on the input of the tooth profile geometry, the dynamic model predicts tooth contact forces, which are passed to the wear model to estimate gear wear and consequently modify the tooth profile geometry, which is then fed back into the dynamic model. This iterative loop enables a knowledge-based prediction of wear and its induced tooth profile changes, which, however, is likely to be reliable only for a limited timeframe, during which the wear model parameters remain unchanged.

The main novelty of the proposed approach relies on the updating of the wear model parameters based on vibration measurements (on the right of Figure 6.1). The vibrations from the gearbox dynamic model are compared to measured vibration

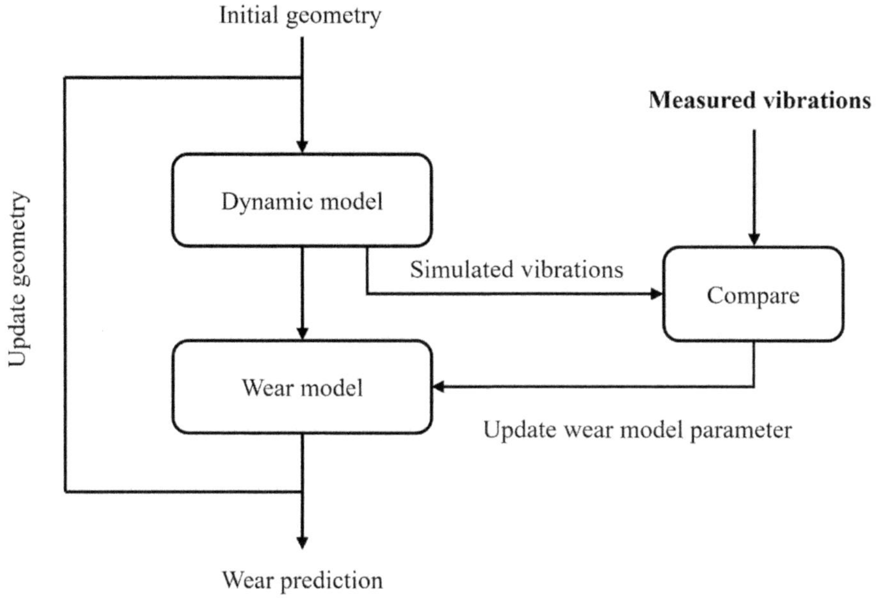

FIGURE 6.1 Proposed vibration-based scheme for updating wear predictions. Made by the author [18].

levels to track the quality of the wear model predictions and, if necessary, update the wear model parameters.

The details of each component of the approach, as outlined in Figure 6.1, will be discussed in detail in the following sections.

6.2.2 Dynamic Model

In this section, only a brief introduction to the dynamic model of the spur gearbox will be given since its development has been introduced with details in Chapter 5.

To generate realistic vibrations and contact forces for the wear model, a 21- DOFs lumped parameter dynamic model is established based on the University of New South Wales (UNSW) gearbox test bench, the development of which has been introduced in Chapter 5. The basic motion equations of the dynamic model are presented as follows:

$$M\ddot{x}+C\dot{x}+Kx=f \tag{6.1}$$

where **x** represents the angular and translational displacements of the different nodes of the gearbox system, which are in the plane perpendicular to the input and output shaft axes. **K**, **C** and **f** are the matrices of corresponding stiffness, damping, and force. Further details of this dynamic model development are provided in Chapter 5. Additionally, the dynamic model was calibrated using impact tests, some speed ramp tests, and several constant speed tests. The calibration ensures that reliable contact force figures are fed into the wear model and that realistic vibration signals are simulated.

6.2.3 Vibration-Based Approach for Wear Mechanism Identification

Although many advanced wear models have been proposed using various methodologies and parameter sets, the Archard wear model remains the most widely used for practical applications, and is chosen in this research to simulate the wear-induced tooth profile change behaviours. Neglecting changes in the contact area over the meshing cycle and modifying the expression from reference [9], the Archard wear model is as follows:

$$\frac{dh}{dt} = K_{wear} F v \tag{6.2}$$

where h is the wear depth, F is the normal load, and v is the corresponding sliding velocity.

Although the capability of the Archard wear model in simulating wear propagation behaviours has been demonstrated in numerous references [19–22], the Archard wear model can be improved to enhance the gear wear prediction accuracy by considering more realistic parameters and factors. First, expressing Eq. (6.2) in terms of contact pressure P rather than contact force would be more physically meaningful [23, 24]. The second point involves the inclusion in the model of the effect from adjacent contact points on the wear depth accumulation at one specific contact point. From the literature review of Chapter 2, it is worth noting that none of the existing works [2, 9, 15, 25–27] considers the pressure contribution from the adjacent contact points to the wear depth accumulation, although the pressure is distributed in a small region and influences the wear distribution.

To address the issues mentioned above, this research further improves the Archard wear model by considering contact pressure rather than force. Additionally, considering the effect of Hertzian deformation and representing the contact area rather than a line, a new approach is proposed to calculate the wear depth distribution:

$$h(x) = \int K_{wear} P(x,t)\, \upsilon(t)\, dt \tag{6.3}$$

where $h(x)$ is the wear depth at contact point x, $P(x,t)$ is the contact pressure distribution at time t, and $\upsilon(t)$ is the corresponding sliding velocity.

To clearly explain the new wear depth distribution calculation approach, a simple example is shown in Figure 6.2 for demonstration. In Figure 6.2, $\mathrm{P}(5)$, $\mathrm{P}(6)$ and $\mathrm{P}(7)$ represent the instantaneous Hertzian pressure distribution at contact points #5, #6, and #7, respectively. The parameter $\mathrm{P}(5)\_6$ represents the consequent pressure at contact point #6 when the gear pairs are engaged at contact point #5, due to the contact area given by Hertzian deformation. A similar meaning holds for $\mathrm{P}(7)\_6$. In summary, due to the Hertzian deformation, the wear depth at contact point #6 is

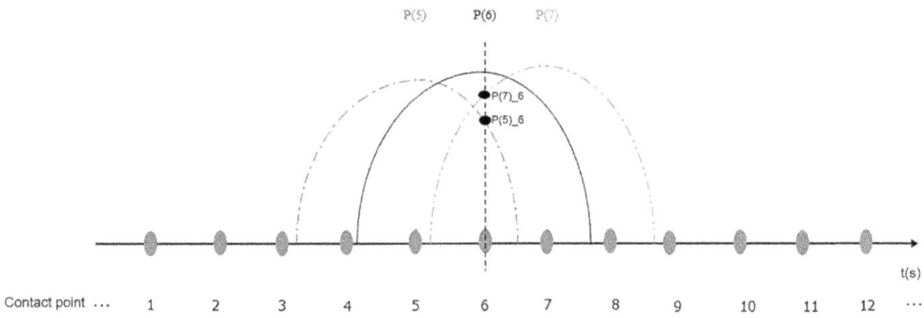

FIGURE 6.2 The wear depth at contact point #6: With effects from adjacent contact points. Made by the author [28].

affected not only when it is the (theoretical) contact 'point', but also when adjacent points #5 and #7 are the (theoretical) contact points. The sum of them forms the final wear depth at contact point #6, as shown in Eq. (6.4).

$$\Delta h(6) = \left(K_{wear} P_6(6) V(6) + K_{wear} P_5(6) V(5) + K_{wear} P_7(6) V(7)\right) \cdot t \qquad (6.4)$$

The gear tooth wear depth distribution obtained using the Archard wear model presented in reference [9] and the improved Archard wear model with the new calculation approach are shown in Figure 6.3. Two differences between the results are evident:

1) the wear depth at the pitch line (rotation angle zero) is non-zero with the new calculation approach, which aligns more closely with experimental observations (see Figure 6.9 in Section 6.3). This is because, although the sliding velocity is theoretically zero at the instant of pitch line contact, due to the Hertzian deformation giving a contact area, the sliding velocity at the pitch line is non-zero during this short instant. This results in mild wear at the pitch line.
2) the curve of the wear depth distribution (from the tooth root to tip) becomes much smoother, which again is more physically meaningful and aligns with experimental observations (see Figure 6.8 in Section 6.3). In practice, although dynamic contact pressure distribution has some peaks corresponding to each contact point (see Figure 6.4), it could not cause sharp peaks and troughs in the wear depth distribution when considering the effect from the adjacent contact points due to Hertzian deformation. Load variations will rapidly remove any sharp peaks and troughs induced by dynamic forces.

Considering these two factors, the wear profile calculated using the improved Archard wear model with the new calculation approach is more reasonable and realistic compared with the Archard wear model used in reference [9]. Therefore, in this

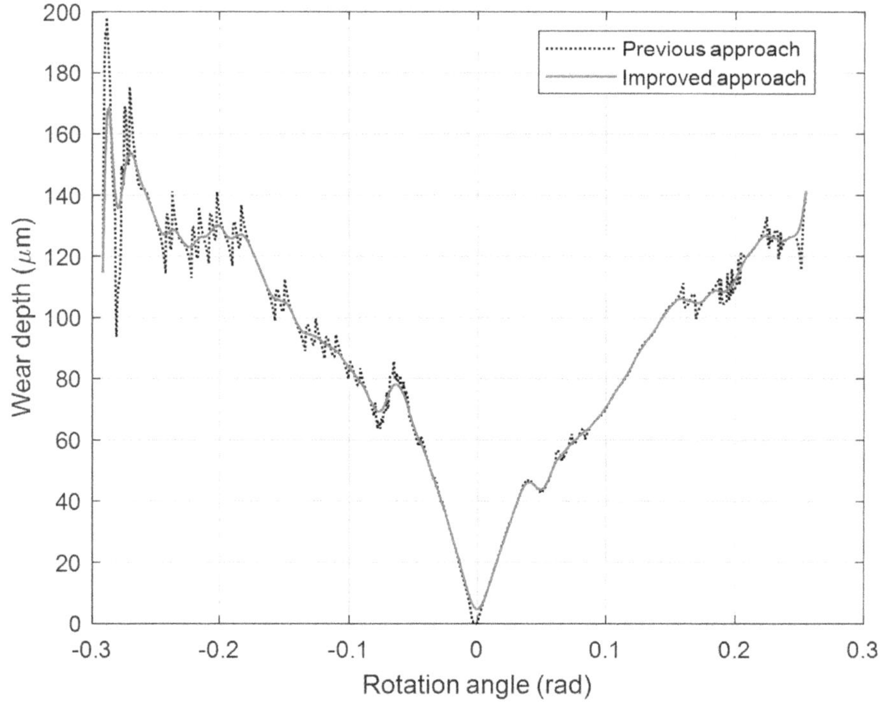

FIGURE 6.3 Wear depth distribution. Made by the author [28].

research, the improved Archard wear model is used to simulate the wear propagation behaviours and its induced tooth profile change.

Note that the worn gear tooth profile (the black dot line) shown in Figure 6.3 is different from the results shown in reference [9]. Specifically, it is not as 'smooth' as the worn tooth profile presented in reference [9]. This is because the single DOF torsional model used in reference [9] neglects some actual modes (such as translational modes) of gear systems, resulting in a simplified and smooth contact force, which does not represent the actual one. In contrast, a 21-DOFs dynamic model is used in this book to account for both translational and torsional modes, resulting in a contact force that differs significantly from the one obtained using the single DOF torsional model. This difference is discussed in Section 5.3. Therefore, the calculated worn gear tooth profile still has small jags, which are caused by the dynamic interactions from translational and torsional modes of the 21-DOFs dynamic model.

It should be noted that the wear model parameter K_{wear} (in Eq. (6.3)) is influenced by several factors including material properties, such as hardness and roughness of the two surfaces, and lubrication conditions, which may evolve in the wear process. Since it is challenging to estimate or directly measure K_{wear} experimentally, this coefficient is often an unknown parameter. In this study, K_{wear} is determined experimentally by comparing simulated and experimental vibrations. The details of this updating methodology will be presented in the following section.

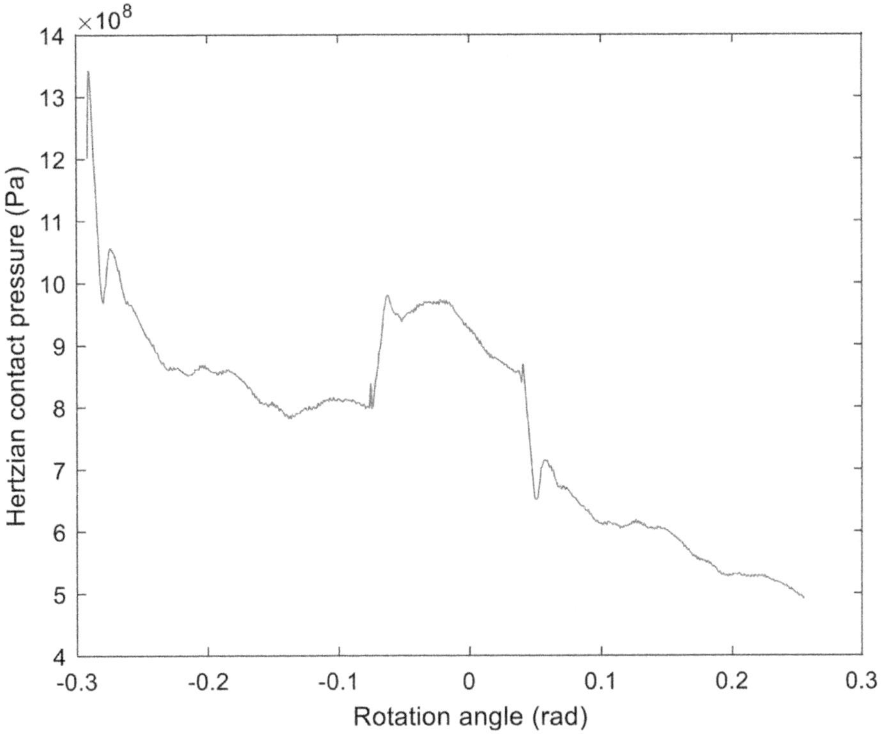

FIGURE 6.4 Contact pressure distribution with peaks at the contact point. Made by the author [28].

6.2.4 Updating Methodology

The wear coefficient K_{wear} is updated by comparing the root-mean-square (RMS) values of simulated and experimental signals. The updating methodology is executed iteratively when the absolute error between the experimental and simulated RMS values exceeds the predefined 5% threshold, and the wear coefficient K_{wear} for the next iteration (i+1) is calculated as follows:

$$K_{\text{wear}(i+1)} = D_{(i)} \cdot K_{\text{wear}(i)} \tag{6.5}$$

where $D_{(i)} = RMS\left\{y^{(EXP,i)}(t)\right\} / RMS\left\{y^{(SIM,i)}(t)\right\}$, and $y^{(EXP,i)}(t)$ and $y^{(SIM,i)}(t)$ denote the experimental and experimental vibration signals at time t and iteration i. Note that, in theory, the predefined threshold should be set close to zero, and the wear coefficient will be updated at nearly every step. Although this can slightly improve prediction accuracy, it significantly increases computational cost. To balance prediction accuracy and computational cost, a 5% threshold is selected in this research. The iterations are stopped when the absolute error between the simulated and experimental RMS values falls below the predefined threshold. It is important to note that the operating conditions of the dynamic model (speed and torque) should be set as close as possible to the actual experimental conditions to avoid potentially large biases on RMS readings. Fortunately, speed measurements and torque estimates (e.g. through current measurements in electromechanical drivetrains) are available on most machines of sufficient criticality to justify a sophisticated condition monitoring system.

6.3 TOOTH PROFILE CHANGE PREDICTION RESULTS

6.3.1 Dry Test

The dry test was conducted with input shaft speed and torque of 10 Hz and5 N·m5 Nm, respectively. The wear model parameter K_{wear} was updated, when necessary, using the procedure discussed in Section 6.2.4. In total, two updates were performed during the test, when the error between the RMS values of simulated and measured vibration signals exceeded the predefined 5% threshold. The first update was performed at 0.73×10^4 wear cycles, and K_{wear} was found to be $1.65 \times 10^{-6}\,Pa^{-1}$. The second update occurred at 2.48×10^4 wear cycles, and the updated K_{wear} was $2.17 \times 10^{-6}\,Pa^{-1}$. Figure 6.5 shows the evolution of the vibration RMS during the dry test. As shown

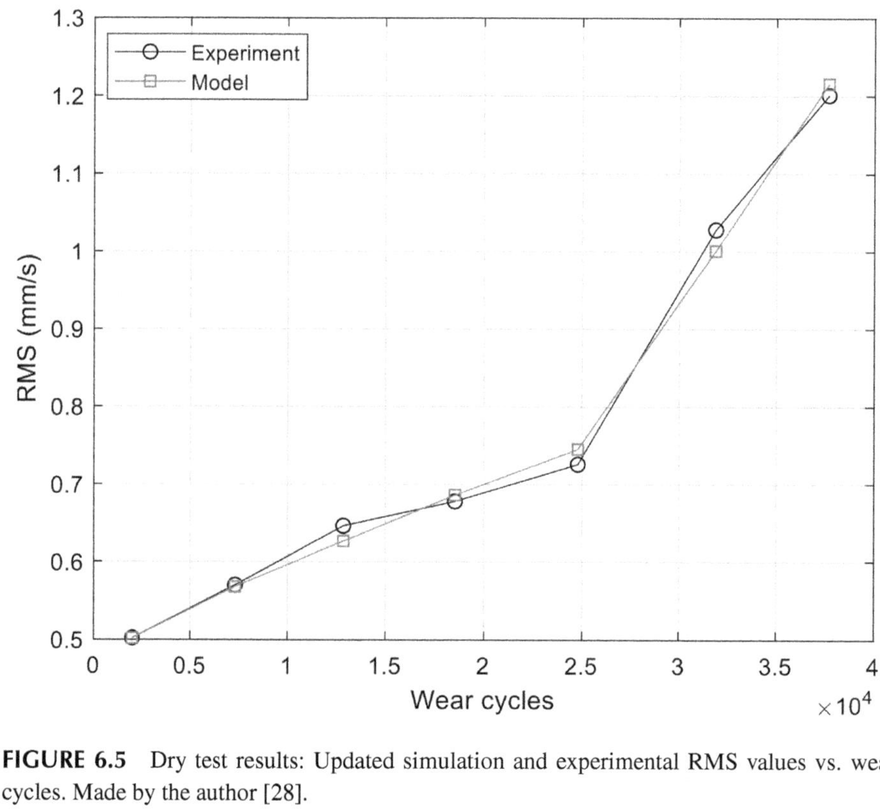

FIGURE 6.5 Dry test results: Updated simulation and experimental RMS values vs. wear cycles. Made by the author [28].

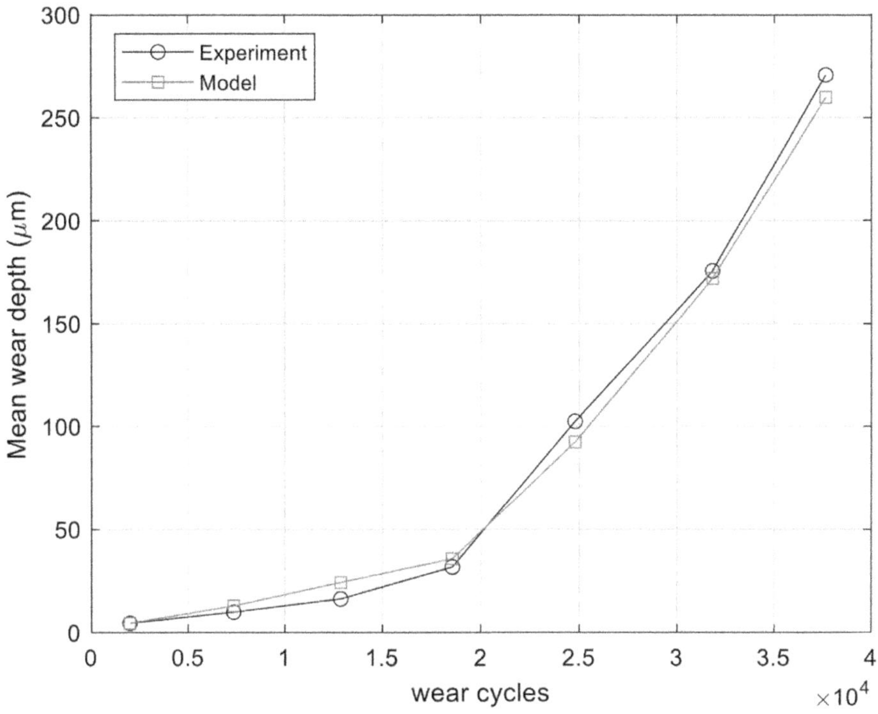

FIGURE 6.6 Wear depth comparison results of dry test: Experiment (from wear particles) and model (after updating). Made by the author [28].

in Figure 6.5, the simulated vibrations from the dynamic model match well with the experimental vibrations after updating K_{wear}.

To assess the effectiveness of the proposed updating scheme (with the improved Archard wear model) in wear depth prediction, an estimate of the mean wear depth was calculated from the wear particles collected using adhesive paper [29]. Figure 6.6 shows the comparison results. As shown in Figure 6.6, the wear depths (at 1.29×10^4, 185×10^4, 3.19×10^4, and 3.77×10^4 wear cycles) are well predicted by the proposed updating scheme. Therefore, the proposed updating scheme with the improved Archard wear model shows excellent performance in wear prediction under dry conditions. Note that this and subsequent mean wear depth figures represent the combined wear of the pinion and driven-gear tooth profiles, which means that the pinion alone accounts for about 73% of this wear depth figure. Note also that these wear depth figures were never used in the updating – only the vibration signals (simulated and measured).

6.3.2 Lubricated Test

This section examines the performance of the wear prediction updating scheme under lubricated conditions. The lubricated test was conducted with input shaft speed and

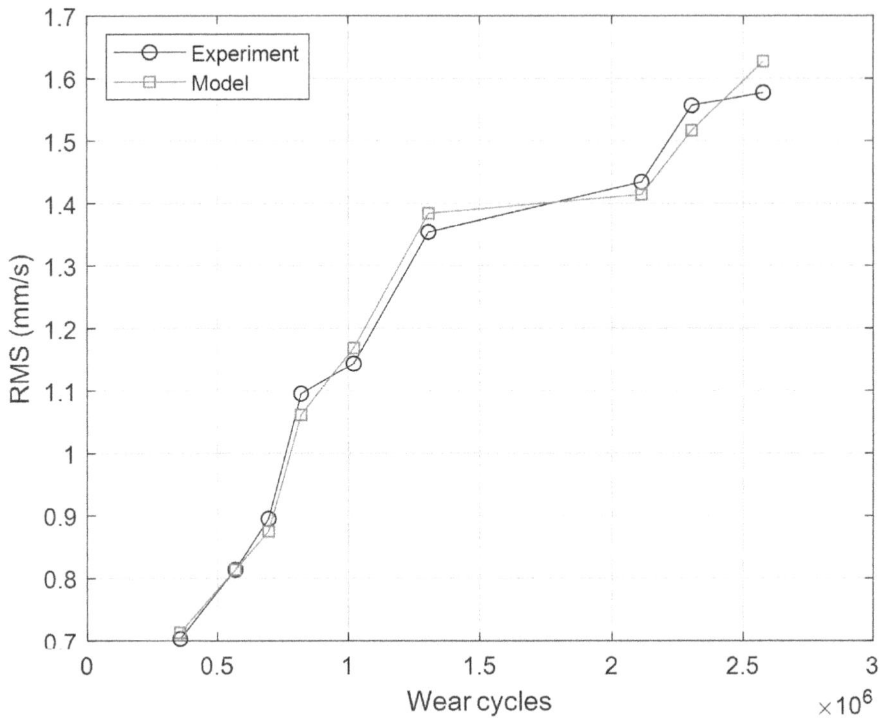

FIGURE 6.7 Testing and the results from simulations (after updating). Made by the author [28].

torque of 20 Hz and 20 N·m, respectively. The wear coefficient, K_{wear}, was updated four times during the test when the error between the RMS values of the simulated and measured vibration signals exceeded a predefined threshold of 5%. The updates occurred at 0.57×10^6, 0.82×10^6, 2.19×10^6, and 2.30×10^6 wear cycles. Figure 6.7 shows the RMS comparison results from the updated model and the experiment. The RMS values from the updated model match well with the experimental ones. It is noted that the manufacturing marks on the driven-gear have been significantly removed during [1.41~2.12] million cycles. The same phenomenon was reported in reference [29]. Thus, a plateau at 1.4 mm/s occurs in the RMS trend during this interval.

In the lubricated test, collecting all wear particles from the oil is difficult. Consequently, the tooth profile change could not be estimated as the dry test using wear particle mass. To acquire reliable wear depth data for verifying the effectiveness of the wear prediction updating scheme, the gear tooth profile was measured experimentally using a moulding technique [29, 30], as shown in Figure 6.8. The tooth profile change from 0.36×10^6 to 2.58×10^6 wear cycles is shown in Figure 6.9. It should be noted that the run-in period (the first 0.36×106 wear cycles) is excluded.

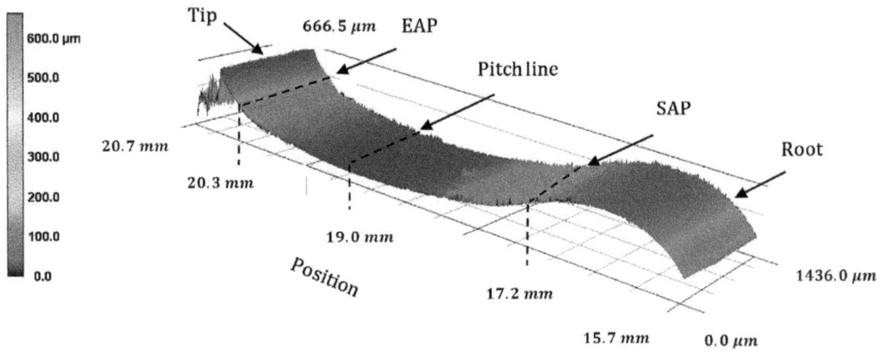

FIGURE 6.8 3D image of the mould profile collected from pinion. Made by the author.

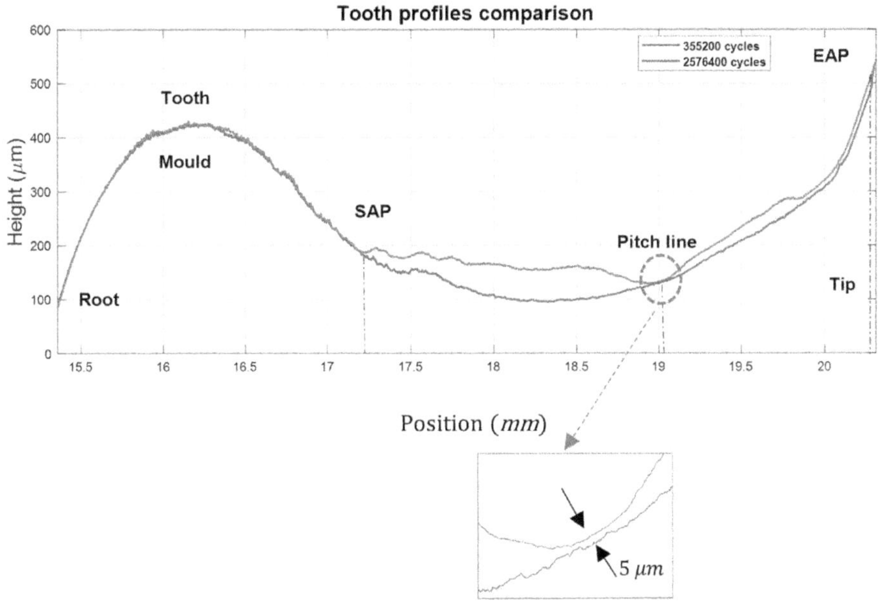

FIGURE 6.9 The tooth profiles of pinion. Made by the author [28].

Based on the profile change demonstrated in Figure 6.9, the mean wear depth (from gear root to tip) of the mating gears should be $48.2\mu m$. The model-based wear depth prediction results compared with the experimental measurement (obtained at the conclusion of the test) are presented in Figure 6.10. From Figure 6.10, it can be found that the final predicted wear depth is $46.2\mu m$. Compared with the measured wear depth, a 4.3% error exists, which is an acceptable discrepancy. Therefore, with the improved Archard wear model, the proposed updating methodology is effective in predicting wear depth under lubricated conditions.

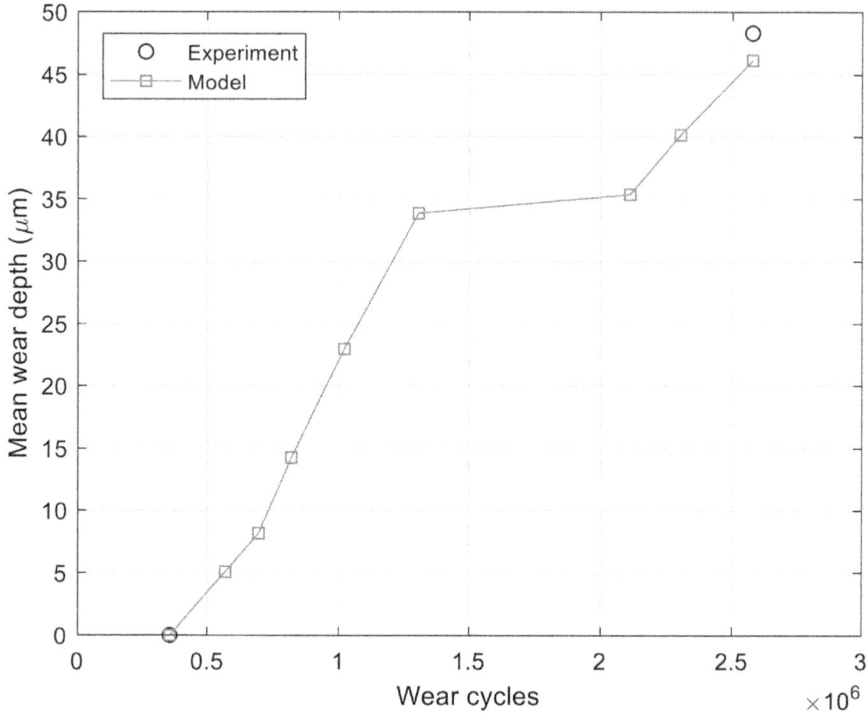

FIGURE 6.10 Wear depth comparison results of the lubricated test. Made by the author [28].

6.4 SUMMARY

The contribution of this chapter is a novel vibration-based updating scheme for monitoring and predicting the gear wear process. In contrast to previous approaches [9, 31, 32], comparison between simulated vibrations and measurements from a physical test rig is conducted regularly to update the model parameters as needed. The updating procedure can track and correct for changes in wear rates, enabling analysts to obtain reliable wear predictions with relatively simple modelling tools. Furthermore, unlike the existing studies in gear wear prediction, the present scheme was applied and validated on both lubricated and dry tests, demonstrating reliable and accurate predictions of the tooth profile change. In addition, unlike reference [25], the method does not require gearbox stoppage or disassembly to measure wear particle mass. By leveraging the data acquisition system, the vibration signals can be easily acquired online without disturbing the operation, allowing continuous application during gearbox operation.

In addition, the developed vibration updating scheme incorporates an improved Archard wear model that calculates wear depth based on pressure distribution instead of localised force, offering a more physically realistic representation. This enhanced model accounts for the effect of Hertzian contact deformation, giving a contact area rather than a line contact. This produces qualitatively more realistic and reliable predictions of wear-induced tooth profile alteration behaviour in comparison with the original approach used in reference [9].

However, it should be mentioned that in the lubricated test, significant fatigue pitting propagation was observed, but the proposed vibration-based updating scheme does not have the capability to predict fatigue pitting propagation. Consequently, Chapter 7 will address this limitation by modifying the updating scheme to simultaneously account for two wear mechanisms (abrasive wear and fatigue pitting) and their respective effects on the gear tooth surface (profile change and surface pitting).

REFERENCES

[1] C. Hu, W.A. Smith, R.B. Randall, Z. Peng, Development of a gear vibration indicator and its application in gear wear monitoring, *Mechanical Systems and Signal Processing*, 76–77 (2016) 319–336. https://doi.org/10.1016/j.ymssp.2016.01.018.

[2] A. Flodin, S. Andersson, Simulation of mild wear in spur gears, *Wear*, 207 (1997) 16–23. https://doi.org/10.1016/S0043-1648(96)07467-4.

[3] Y. Huangfu, Z. Zhao, H. Ma, H. Han, K. Chen, Effects of tooth modifications on the dynamic characteristics of thin-rimmed gears under surface wear, *Mechanism and Machine Theory*, 150 (2020) 103870. https://doi.org/10.1016/j.mechmachtheory.2020.103870.

[4] I. Yesilyurt, F. Gu, A.D. Ball, Gear tooth stiffness reduction measurement using modal analysis and its use in wear fault severity assessment of spur gears, *Ndt & E International*, 36 (2003) 357–372. https://doi.org/10.1016/S0963-8695(03)00011-2.

[5] C. Yuksel, A. Kahraman, Dynamic tooth loads of planetary gear sets having tooth profile wear, *Mechanism and Machine Theory*, 39 (2004) 695–715. https://doi.org/10.1016/j.mechmachtheory.2004.03.001.

[6] J. Wojnarowski, V. Onishchenko, Tooth wear effects on spur gear dynamics, *Mechanism and Machine Theory*, 38 (2003) 161–178. https://doi.org/10.1016/S0094-114X(02)00091-5.

[7] F.K. Choy, V. Polyshchuk, J.J. Zakrajsek, R.F. Handschuh, D.P. Townsend, Analysis of the effects of surface pitting and wear on the vibration of a gear transmission system, *Tribology International*, 29 (1996) 77–83. https://doi.org/10.1016/0301-679X(95)00037-5.

[8] J.H. Kuang, A.D. Lin, The effect of tooth wear on the vibration spectrum of a spur gear pair, *Journal of Vibration and Acoustics*, 123 (2001) 311–317. https://doi.org/10.1115/1.1379371.

[9] H. Ding, A. Kahraman, Interactions between nonlinear spur gear dynamics and surface wear, *Journal of Sound and Vibration*, 307 (2007) 662–679. https://doi.org/10.1016/j.jsv.2007.06.030.

[10] A. Kahraman, H. Ding, A methodology to predict surface wear of planetary gears under dynamic conditions, *Mechanics Based Design of Structures and Machines*, 38 (2010) 493–515. https://doi.org/10.1080/15397734.2010.501312.

[11] X. Liu, Y. Yang, J. Zhang, Investigation on coupling effects between surface wear and dynamics in a spur gear system, *Tribology International*, 101 (2016) 383–394. https://doi.org/10.1016/j.triboint.2016.05.006.

[12] Z. Shen, B. Qiao, L. Yang, W. Luo, Z. Yang, X. Chen, Fault mechanism and dynamic modeling of planetary gear with gear wear, *Mechanism and Machine Theory*, 155 (2021) 104098. https://doi.org/10.1016/j.mechmachtheory.2020.104098.

[13] B. Hu, C. Zhou, H. Wang, S. Chen, Nonlinear tribo-dynamic model and experimental verification of a spur gear drive under loss-of-lubrication condition, *Mechanical Systems and Signal Processing*, 153 (2021) 107509. https://doi.org/10.1016/j.ymssp.2020.107509.

[14] W. Chen, Y. Lei, Y. Fu, L. Hou, A study of effects of tooth surface wear on time-varying mesh stiffness of external spur gear considering wear evolution process, *Mechanism and Machine Theory*, 155 (2021) 104055. https://doi.org/10.1016/j.mechmachtheory.2020.104055.

[15] P. Bajpai, A. Kahraman, N.E. Anderson, A surface wear prediction methodology for parallel-axis gear pairs, *Journal of Tribology*, 126 (2004) 597–605. https://doi.org/10.1115/1.1691433.

[16] M. Masjedi, M.M. Khonsari, On the prediction of steady-state wear rate in spur gears, *Wear*, 342–343 (2015) 234–243. https://doi.org/10.1016/j.wear.2015.08.010.

[17] W.D. Mark, *Performance-based gear metrology*, Wiley Online Library, 2012.

[18] K. Feng, P. Borghesani, W.A. Smith, R.B. Randall, Z.Y. Chin, J. Ren, Z. Peng, Vibration-based updating of wear prediction for spur gears, *Wear*, 426–427 (2019) 1410–1415. https://doi.org/10.1016/j.wear.2019.01.017.

[19] A. Kahraman, P. Bajpai, N.E. Anderson, Influence of tooth profile deviations on helical gear wear, *Journal of Mechanical Design*, 127 (2004) 656–663. https://doi.org/10.1115/1.1899688.

[20] T.L. Krantz, A. Kahraman, An experimental investigation of the influence of the lubricant viscosity and additives on gear wear, *Tribology Transactions*, 47 (2004) 138–148. https://doi.org/10.1080/05698190490278949.

[21] M.Ş. Tunalioğlu, B. Tuç, Theoretical and experimental investigation of wear in internal gears, *Wear*, 309 (2014) 208–215. https://doi.org/10.1016/j.wear.2013.11.016.

[22] K. Mao, W. Li, C.J. Hooke, D. Walton, Friction and wear behaviour of acetal and nylon gears, *Wear*, 267 (2009) 639–645. https://doi.org/10.1016/j.wear.2008.10.005.

[23] J. Andersson, A. Almqvist, R. Larsson, Numerical simulation of a wear experiment, *Wear*, 271 (2011) 2947–2952. https://doi.org/10.1016/j.wear.2011.06.018.

[24] D. Park, A. Kahraman, A surface wear model for hypoid gear pairs, *Wear*, 267 (2009) 1595–1604. https://doi.org/10.1016/j.wear.2009.06.017.

[25] F. Zhao, Z. Tian, X. Liang, M. Xie, An integrated prognostics method for failure time prediction of gears subject to the surface wear failure mode, *IEEE Transactions on Reliability*, 67 (2018) 316–327. https://doi.org/10.1109/TR.2017.2781147.

[26] T. Osman, P. Velex, Static and dynamic simulations of mild abrasive wear in wide-faced solid spur and helical gears, *Mechanism and Machine Theory*, 45 (2010) 911–924. https://doi.org/10.1016/j.mechmachtheory.2010.01.003.

[27] Z. Shen, B. Qiao, L. Yang, W. Luo, X. Chen, Evaluating the influence of tooth surface wear on TVMS of planetary gear set, *Mechanism and Machine Theory*, 136 (2019) 206–223. https://doi.org/10.1016/j.mechmachtheory.2019.03.014.

[28] K. Feng, W.A. Smith, Z. Peng, Use of an improved vibration-based updating methodology for gear wear prediction, *Engineering Failure Analysis*, 120 (2021) 105066. https://doi.org/10.1016/j.engfailanal.2020.105066.

[29] Z.Y. Chin, W.A. Smith, P. Borghesani, R.B. Randall, Z. Peng, Absolute transmission error: A simple new tool for assessing gear wear, *Mechanical Systems and Signal Processing*, 146 (2021) 107070. https://doi.org/10.1016/j.ymssp.2020.107070.

[30] H. Chang, P. Borghesani, W.A. Smith, Z. Peng, Application of surface replication combined with image analysis to investigate wear evolution on gear teeth – A case study, *Wear*, 430–431 (2019) 355–368. https://doi.org/10.1016/j.wear.2019.05.024.

[31] D. Park, M. Kolivand, A. Kahraman, Prediction of surface wear of hypoid gears using a semi-analytical contact model, *Mechanism and Machine Theory*, 52 (2012) 180–194. https://doi.org/10.1016/j.mechmachtheory.2012.01.019.

[32] Z. Zhang, C. Zheng, M. Wen, S. Yang, H. Li, Q. Du, *A wear prediction model for spur gears based on the dynamic meshing force and tooth profile reconstruction*, 2nd International Conference on Computer Engineering, Information Science & Application Technology (ICCIA 2017), Atlantis Press, 2016.

7 Digital Twin for Gear Pitting and Profile Change Monitoring and Prediction

7.1 INTRODUCTION

In gear systems, wear-induced gear tooth surface degradation is an inevitable process, and it can result in destructive tooth damage and a substantial reduction in the remaining useful life (RUL) of the gearbox [1–4]. Consequently, monitoring and predicting wear propagation is essential for scheduling timely maintenance and preventing catastrophic failures. Tooth profile changes and surface pitting of gear teeth are two predominant forms during gear service life, and they have different impacts on the degradation rate and RUL of gear systems. During the gear wear process, these two wear mechanisms can occur concurrently, and they can act together, leading to a more rapid surface deterioration than either would in isolation [5–8]. Therefore, the development of an efficient and reliable tool capable of monitoring and predicting both gear tooth profile changes and surface pitting could bring enormous benefits to the industry.

Based on the published research in Chapter 2, the Archard wear model [9–11] is widely employed for estimating the wear depth of tooth surfaces. By integrating the Archard wear model, dynamic model, and measured vibrations, the wear-induced tooth profile change under dynamic conditions can be predicted as presented in Chapter 6. As reviewed in Chapter 2, compared with research on the prediction of abrasive wear and associated tooth profile changes, studies on surface pitting propagation monitoring and predicting are sparser, despite extensive studies on pitting initiation mechanisms [12–14]. References [15] and [5] have employed the Archard wear model, the Dang Van fatigue criterion [16–18], and the Lundberg-Palmgren model [19–22] to simulate surface pitting propagation and investigate the coupling effects between abrasion and pitting. Although these studies demonstrated promising surface pitting prediction results, it should be noted that neither study included model calibration against actual measurements to account for variations in wear rate. In practice, the pitting propagation rate is influenced by a number of factors, such as lubrication quality and quantity, contact pressure distribution, sliding speed, and surface roughness, all of which may change significantly. Without frequent checking and, if necessary, updating of the wear model parameters, prediction accuracy cannot be guaranteed and is likely to decrease significantly during the surface pitting propagation.

DOI: 10.1201/9781003686736-7

In summary, existing methods lack the capability for accurate prediction of gear surface degradation, especially during the concurrent occurrence of tooth profile change and surface pitting. This limitation necessitates the development of a reliable and efficient tool, which can utilise the capability of both physics-based models and measurements for predicting both tooth profile change and surface pitting propagation.

A two-step approach is used to achieve the above goal. The first step involves a method for monitoring and predicting gear tooth profile change in Chapter 6. This methodology is based on the combination of an Archard wear model, a dynamic model, and an updating scheme that recalibrates the Archard wear coefficient as required. By comparing simulated and measured vibrations, the method reliably predicts abrasion-induced tooth profile changes. However, this methodology does not have the ability to monitor and predict surface pitting behaviours. Furthermore, the root mean square (RMS) of the raw vibration signal was used as the feature in the comparative analysis. However, this feature is susceptible to contamination from background noise and unrelated machine dynamics, which can severely degrade prediction accuracy and increase computational cost.

Built upon the work introduced in Chapter 6, this chapter presents a vibration-based methodology for monitoring and predicting both wear-induced tooth profile change and surface pitting.

7.2 RELATIONSHIPS OF VIBRATION FEATURES AND WEAR FEATURES

In most applications, with constant or randomly varying speeds and loads, tooth profile change and surface pitting tend to be distributed across all gear teeth uniformly. But these two wear mechanisms have different impacts on gear systems [23], and thus result in distinct vibration characteristics. Tooth profile change generally increases the amplitude of gear mesh harmonics [1, 24]. Consequently, this phenomenon can be effectively monitored by the RMS of the signal obtained by synchronously averaging over the gear mesh period (giving a signal comprising only gear mesh harmonics), as defined in Eq. (7.1):

$$RMS_{SA} = \sqrt{\sum_{i=1}^{N} A_{GMi}^2} \tag{7.1}$$

where A_{GMi} is the amplitude of the ith gear mesh harmonic. The merits of using the time-synchronously averaged (TSA) vibration signal are as follows: 1) Environmental noise can be attenuated significantly; 2) the average level of profile change of all gear teeth can be well indicated. It should be noted that well-lubricated gear systems usually exhibit relatively smooth worn tooth profiles characterised by long-wavelength variations from the involute. For such mild wear cases, the first few gear mesh harmonics are sufficient to capture the extent of profile change. Conversely, the severely worn tooth profiles that are more often found in dry or poorly lubricated conditions are more complex, and therefore, more gear mesh harmonics are required to properly reflect such profiles. This difference will be demonstrated with a lubricated test and a dry test in the results section.

As for surface pitting, it is fatigue-induced material loss. Following its occurrence, the effective working tooth profile across the entire face width typically remains unchanged, except that pitting is extremely severe. Consequently, the gear mesh harmonics often exhibit negligible change during this process. To monitor the surface pitting propagation, ICS2, a measure of the degree of second-order cyclostationarity in a signal [25] and shown in Eq. (7.2), was investigated for tracking surface pitting propagation in Chapter 4, considering of the underlying physics of the gear meshing process:

$$\text{ICS2}^{\mathcal{A}_h,\text{H}} = \frac{\sum_{\text{h}=1:\text{H}} \max_{\text{n}\in\mathcal{A}_h} \left(\text{SES}[\text{n}]^2\right)}{\text{SES}[0]^2} \tag{7.2}$$

where h is the harmonic order of the gear mesh frequency, and $\mathcal{A}_h$ ($h = 1, 2, \ldots,$ H) represents the equivalent sets for the corresponding harmonics. A tolerance band in the case of expected cyclic frequency deviations, to account for imperfect order tracking for instance. SES is the (amplitude) spectrum of the squared envelope, obtained through amplitude demodulation of the signal. Using a bandpass filtered version of the signal to compute the SES enhances the ICS2 metric by targeting not just the desired cyclic frequencies but also the desired carrier frequency range. This form of ICS2 has proven successful in difficult bearing diagnostic cases [26], and the experimental results presented in Chapter 4 further confirm that ICS2 has an excellent performance in tracking surface pitting propagation when based on a low carrier frequency band. Therefore, this book employs ICS2 based on a low-frequency band as the vibration feature to conduct the comparison analysis for surface pitting propagation monitoring and prediction, with details provided in Section 7.3

7.3 METHODOLOGY FOR MONITORING AND PREDICTING SURFACE PITTING AND TOOTH PROFILE CHANGE

This section presents the proposed vibration-based surface degradation monitoring and prediction methodology. It details the development of the dynamic and tribological/wear models, followed by the approaches of model updating using measured vibrations.

7.3.1 Structure of the Proposed Vibration-Based Surface Degradation Prediction Methodology

The architecture of the proposed methodology is illustrated in Figure 7.1. It comprises a dynamic model, wear models (including the Archard wear model and a surface pitting model), and comparison analysis with measured vibrations for updating wear coefficients.

A dynamic model is established to simulate the actual spur gearbox. Since the dynamic model development was detailed in Chapter 5, it will not be introduced in this chapter to avoid repetition. Based on the inputs of gear tooth profile geometry and operating conditions, the dynamic model can provide dynamic contact forces and vibration responses of the gear system. The contact pressure can then be calculated using the Hertzian contact theory. With the contact pressure as an input, the Archard wear model can estimate the abrasive wear rate and consequently predict the gear tooth profile geometry at a specified future point. The updated profile is then fed back into the dynamic model. Simultaneously, the contact pressure is input to a surface pitting model to predict the surface pitting density, which consequently modifies the contact area thereby alters the Hertzian contact pressure. Further information about the wear models can be provided in Section 7.3.2. Both the profile change and surface pitting can affect the contact pressure, through modifying gear tooth profile geometry and contact area, respectively. Therefore, there is an interaction between tooth profile change and surface pitting propagation.

Within the proposed methodology, the contact pressure from the dynamic model serves as input to the two wear models (Archard wear model and surface pitting model), enabling simultaneous prediction of wear depth and surface pitting density. In Figure 7.1, this part of the methodology is represented mostly in black. The blue elements represent the updating of the contact area and tooth profile geometry, both of which can be performed as often as required without additional measurements.

In practice, numerous factors can affect the dominant wear mechanisms and wear rate, such as the quality of lubrication and changes in the surface roughness and operating conditions. Consequently, employing a constant wear coefficient K in each wear model could lead to significant prediction errors. To ensure accuracy, regular updating of the wear coefficients is implemented by comparing available measured vibrations (RMS_{SA} and ICS2 in Section 7.2) with the simulations. This updating procedure, illustrated in Figure 7.1, the RMS_{SA} of the measured and simulated vibrations is compared directly. In contrast, the ICS2 is not compared directly. Instead, ICS2 from the measurements is used to estimate pitting density, which is then compared with the simulation. This is because accurately modelling the vibrational effect of fatigue pitting is highly challenging; it requires an overly complex contact model and a highly detailed dynamic model to replicate the vibration responses across the desired frequency range.

In summary, once the simulations diverge from actual measurements, two separate updating procedures are executed on the wear coefficients in the two models. The approaches for updating wear coefficients will be introduced in Section 7.3.3.

7.3.2 Wear Models: Modelling Surface Pitting Behaviours

As the Archard wear model and its enhancements for simulating wear-induced tooth profile change were presented in Chapter 6, this section focuses exclusively on the newly developed surface pitting model.

The Lundberg-Palmgren fatigue model (Eq. (7.3)) proposed in reference [19] is a widely adopted theory for determining the fatigue life of rolling element bearings, with its advantages including excellent performance and high computational efficiency. It

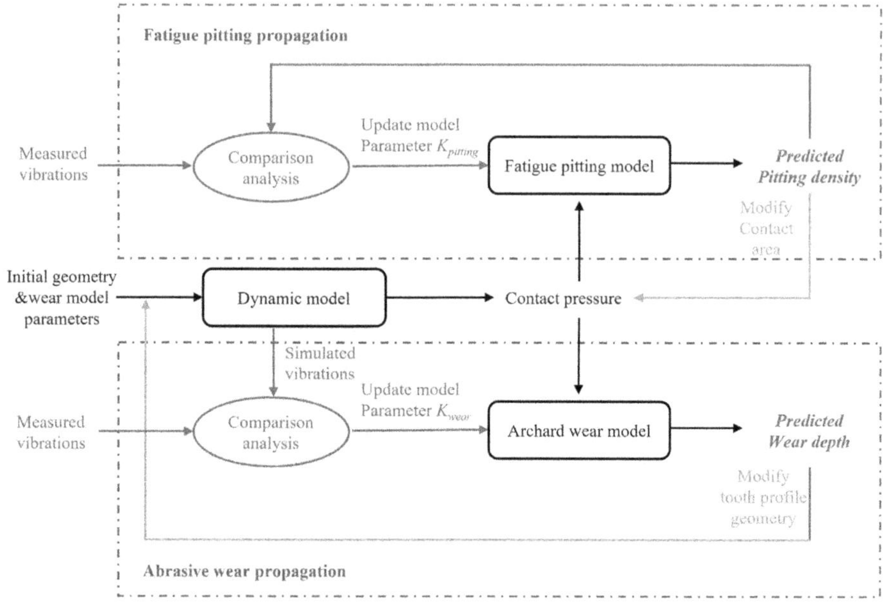

FIGURE 7.1 Basic procedure of vibration-based updating methodology for gear wear prediction. Made by the author [27].

has also been extended to gear systems [28–30] for estimating the RUL when surface pitting propagates. It can be defined as:

$$ln\frac{1}{S(N,\theta)} \sim \frac{\tau_0(\theta)^c N^m}{z_0(\theta)^h} V(\theta) \tag{7.3}$$

where $S(N,\theta)$ denotes the survival probability of the gear tooth at specific rotation angle θ after N running cycles, and c, m and h are material coefficients derived from gear material properties [31]. Some parameters are determined through Hertzian contact theory, including maximum shear stress τ_0, stressed volume V, and depth of the maximum shear stress z_0 [31–33], defined as:

$$\tau_0(\theta) = 0.3 \times P_{max}(\theta) \tag{7.4}$$

$$z_0(\theta) = 0.786 \times b(\theta) \tag{7.5}$$

$$V(\theta) = 2b(\theta) \times z_0(\theta) \times 2\pi \times R(\theta) \tag{7.6}$$

where P_{max} is the maximum Hertzian contact pressure at the specific rotation angle θ, and b and R represent the Hertzian contact radius and equivalent radius, respectively.

However, the published researches in Chapter 2 indicate that the model has not been developed for simulating pitting propagation behaviours. To address this issue, recently published research [5] equating the level of surface pitting damage to the reciprocal of S. , and modelling the distribution of fatigue pitting on the gear tooth, but its severity was not simulated or assessed. Inspired by this research, and furthermore considering that the range of S is $[0,1]$, we propose instead to approximate surface pitting density as:

$$D(N,\theta) \sim 1 - S(N,\theta) \tag{7.7}$$

With this assumption, the fatigue pitting propagation behaviour can be simulated and predicted. However, two issues remain with respect to Eq. (7.7):

1) the pitting propagation behaviour is determined by the initial contact stresses, and neglecting any changes in these stresses due to abrasive wear.
2) there is no feedback mechanism to account for the fact that prior pitting significantly affects the pitting propagation rate.

In practice, the pitting propagation rate would be affected by multiple factors, such as lubrication quality and quantity, contact pressure distribution, surface roughness, and operating conditions. Additionally, the previous pitting density would indeed influence the rate and nature of further pitting propagation. To address these issues, $K_{pitting}$, the surface pitting coefficient, is introduced to update the surface pitting propagation rate based on actual measurements. The surface pitting propagation model in Eq. (7.7) is thus further extended to Eq. (7.8):

$$D(N,\theta) = 1 - \frac{1}{\exp\left(\frac{K_{pitting} \cdot \tau_0(\theta)^c \cdot N^m \cdot V(\theta)}{z_0(\theta)^h}\right)} \tag{7.8}$$

To realise the accumulated pitting propagation behaviour, the surface pitting propagation rate $D'(N,\theta)$ is calculated as follows:

$$D'(N,\theta) = \frac{dD(N,\theta)}{dN}$$

$$= \frac{K_{pitting} \times \tau_0(\theta)^c \times m \times V(\theta) \times N^{m-1} \times \exp\left(-\frac{K_{pitting} \times \tau_0(\theta)^c \times V(\theta) \times N^m}{z_0(\theta)^h}\right)}{z_0(\theta)^h} \tag{7.9}$$

The surface pitting increment ΔD during ΔN running cycles is shown in Eq. (7.10):

$$\Delta D(\Delta N,\theta) = \int_{N}^{N+\Delta N} D'(N,\theta)\,dN$$

$$= \exp\left(-\frac{K_{\text{pitting}} \times \tau_0(\theta)^c \times V(\theta) \times N^m}{z_0(\theta)^h}\right)$$

$$-\exp\left(-\frac{K_{\text{pitting}} \times \tau_0(\theta)^c \times V(\theta) \times (N+\Delta N)^m}{z_0(\theta)^h}\right) \tag{7.10}$$

The accumulated pitting density D_{i+1} at $N+\Delta N$ running cycles is:

$$D_{i+1} = D_i + (\exp\left(-\frac{K_{\text{pitting}} \times \tau_0(\theta)^c \times V(\theta) \times N^m}{z_0(\theta)^h}\right)$$

$$-\exp\left(-\frac{K_{\text{pitting}} \times \tau_0(\theta)^c \times V(\theta) \times (N+\Delta N)^m}{z_0(\theta)^h}\right) \tag{7.11}$$

where D_i denotes the pitting density at N running cycles. Note that $\tau_0(\theta)$, $V(\theta)$ and $z_0(\theta)$ will be updated based on the pitting density D_i, which will be introduced below.

The surface pitting can reduce the actual contact length (B), and then affect further surface pitting propagation, which is defined in Eq. (7.12):

$$B(N_{i+1}) = B_{\text{initial}} \times (1 - D(N_i)) \tag{7.12}$$

where B_{initial} is the designed gear tooth width.

Then $\tau_0(N_{i+1})$, $V(N_{i+1})$ and $z_0(N_{i+1})$ will be calculated with the previous pitting density $D(N_i)$ based on Hertzian theory, and the pitting density $D(N_{i+1})$ can be obtained.

Eqs (7.11) and (7.12) enable the simulation of the accumulated pitting propagation process. Using accumulated pitting density as in Eq. (7.11), in which τ_0, V and z_0 are functions of the running cycles N, allows for the fact that existing pitting (and indeed any existing changes in the tooth profile from abrasive wear) would definitely promote and affect future pitting propagation. This feedback effect of both abrasive wear and fatigue pitting can be conducted without vibration measurements, and is represented by the blue parts shown in Figure 7.1.

7.3.3 Model Updating Procedures Using Measured Vibrations

As indicated by the purple part shown in Figure 7.1, to deliver an accurate wear prediction result in the proposed methodology, measured vibration signals are used to determine whether any updating of the wear model coefficient K is required. This section details the model updating procedures for gear tooth profile change and surface pitting propagation.

To determine whether the wear coefficient K_{wear} in Archard's model requires updating, the simulated and measured vibration signals are compared as introduced in Chapter 6. The updating methodology is executed iteratively when the absolute error between the simulated and experimental RMS_{SA} values surpasses the predefined 5% threshold. The wear coefficient for the next iteration is calculated as follows:

$$K_{\text{wear}(j+1)} = G_{(j)}^{-1} \cdot K_{\text{wear}(j)} \tag{7.13}$$

where

$$G_{(j)} = RMS_{SA}\left\{y^{(\text{SIM},j)}(t)\right\} / RMS_{SA}\left\{y^{(\text{EXP},j)}(t)\right\} \tag{7.14}$$

in which, $y^{(\text{SIM})}(t)$ and $y^{(\text{EXP})}(t)$ denote the simulated and experimental vibration signals. This iterative loop permits knowledge-based monitoring and prediction of the tooth profile change. Note that the index j, associated with running cycles here, is distinct from the index i used in Eqs (7.11) and (7.12). While i corresponds to the frequency of 'blue loop' updates shown in Figure 7.1 – the wear feedback effects contained entirely within the simulation model – which can be executed arbitrarily often, j corresponds to the frequency of 'purple loop' updates, which are executed based on measurement availability.

Since pitting severity can be estimated based on instantaneous measurements, the pitting coefficient K_{pitting} can be periodically updated to account for changes in the pitting growth rate, which may result from lubrication contamination or surface contact temperature changes. In the following, the updating procedure will be introduced.

The ICS2 of measured vibrations in the low frequency region has demonstrated excellent performance in tracking surface pitting propagation in Chapter 4, and is therefore employed here as the vibration feature for the updating procedure. If the surface pitting model performs well, the predicted pitting density should also have a high correlation with ICS2. Otherwise, the surface pitting model parameter K_{pitting} should be updated to improve pitting prediction accuracy. Note that although a strong correlation between low-frequency ICS2 and pitting density (area) was found in Chapter 4, the 'scale factor' is still missing, meaning the actual pitting level cannot be reliably determined through direct measurements. Consequently, at least one detailed inspection of the gear surface is necessary to establish this 'scale factor,' which will be introduced in the following procedure.

Chapter 4 established a linear relationship between measured pitting density and ICS2 of measured vibrations, quantified by the coefficient of determination R^2. For a single regressor, fitted by least squares, the coefficient of determination R^2 equals the square of Pearson's correlation coefficient C [34]. Thanks to its clear analytical expression and simple computational procedure, Pearson's correlation coefficient [35] is employed here to determine the optimal K_{pitting}, ensuring a strong correlation between the ICS2 of measured vibrations and predicted pitting density.

The predicted pitting density array can be defined as $\mathbf{X} = \left[D_{j-1}, D_j, D_{j+1}\right]$ and measured ICS2 array as $\mathbf{Y} = \left[ICS_{j-1}, ICS2_j, ICS2_{j+1}\right]$. Taking $\overline{x} = \frac{D_{j-1} + D_j + D_{j+1}}{3}$ and $\overline{y} = \frac{ICS_{j-1} + ICS2_j + ICS2_{j+1}}{3}$, the correlation coefficient C between X and Y is given by:

$$C = \frac{\left(D_{j-1} - \overline{x}\right)\left(ICS2_{j-1} - \overline{y}\right) + \left(D_j - \overline{x}\right)\left(ICS2_j - \overline{y}\right) + \left(D_{j+1} - \overline{x}\right)\left(ICS2_{j+1} - \overline{y}\right)}{\sqrt{\left(D_{j-1} - \overline{x}\right)^2 + \left(D_j - \overline{x}\right)^2 + \left(D_{j+1} - \overline{x}\right)^2}\sqrt{\left(ICS2_{j-1} - \overline{y}\right)^2 + \left(ICS2_j - \overline{y}\right)^2 + \left(ICS2_{j+1} - \overline{y}\right)^2}} \tag{7.15}$$

Now define correlation coefficient C as a function of K_{pitting}:

$$C\left(K_{\text{pitting}}\right) = f\left(K_{\text{pitting}}\right), K_{\text{pitting}} \in \left[0, +\infty\right) \tag{7.16}$$

If the surface pitting model needs to be updated due to changes in the pitting propagation rate, the updated $K_{\text{pitting}\,\textbf{updated}}$ should maximise the correlation coefficient C, at which point the derivative of C should be 0:

$$C'\left(K_{\text{pitting}\,\textbf{updated}}\right) = 0 \tag{7.17}$$

The correlation analysis utilises only three ICS2_j values and predicted surface pitting densities D_j, ensuring the correlation coefficient C is highly sensitive to changes in the pitting growth rate so that the optimal pitting coefficient K_{pitting} can be obtained.

The detailed updating procedure for K_{pitting} is summarised as follows:

step 1: Measure ICS2 to identify the specific pitting occurrence time and define it as $j = 0$. Note: after the run-in period, surface pitting propagation typically increases the ICS2 of measured vibrations as presented in Chapter 4.

step 2: Use a mould image [36] if possible (or visual inspection) to obtain the actual pitted density: $D_{\text{actual}}\left(N_0\right)$ and $D_{\text{actual}}\left(N_1\right)$.

step 3: The procedures for determining the simulated pitting density $D\left(N_2\right)$ and predicting the pitting density $D\left(N_3\right)$ are as follows:

1) build vibration reference $\left[ICS2_0, ICS2_1, ICS2_2\right]$.
2) use surface pitting model to obtain the simulated pitting density $D(N_2)$ with an initial pitting coefficient $K_{\text{pitting}}(2)$.
3) if Pearson's correlation coefficient C between $\left[ICS2_0, ICS2_1, ICS2_2\right]$ and $\left[D_{\text{actual}}(N_0), D_{\text{actual}}(N_1), D(N_2)\right]$ exceeds 0.95, it means the surface pitting model at time $j = 2$ performs well, and $K_{\text{pitting}}(3) = K_{\text{pitting}}(2)$ can be used to predict surface pitting density $D(N_3)$ at time $j = 3$.
4) if Pearson's correlation coefficient C is below 0.95, it means the simulated surface pitting density $D(N_2)$ is not correct, and the surface pitting coefficient needs to be updated to obtain $K_{\text{pitting}}(2)$.

Using the above-mentioned updating approach in Eq. (7.17), the optimal $K_{\text{pitting}}(2)$ can be determined (then obtain updated $D(N_2)$), which ensures a strong correlation between $\left[ICS2_0, ICS2_1, ICS2_2\right]$ and $\left[D_{actual}(N_0), D_{actual}(N_1), D(N_2)\right]$. After that, set $K_{\text{pitting}}(3) = K_{\text{pitting}}(2)_{\textbf{updated}}$ to predict pitting density $D(N_3)$ at time $j = 3$;

step 4: Repeat the above-mentioned process. At time j, current surface pitting coefficient $K_{\text{pitting}}(j) = K_{\text{pitting}}(j-1)$, and the predicted pitting density is $D(N_j)$. If the correlation coefficient C between $\left[D(N_{j-2}), D(N_{j-1}), D(N_j)\right]$ and $\left[ICS2_{j-2}, ICS2_{j-1}, ICS2_j\right]$ exceeds 0.95, it means the current predicted pitting density $D(N_j)$ is correct, and then set $K_{\text{pitting}}(j+1) = K_{\text{pitting}}(j)$ to predict the pitting density $D(N_{j+1})$ at time $j+1$.

Otherwise, the above-mentioned updating approach in Eq. (7.17) is applied to obtain the updated pitting coefficient $K_{\text{pitting}}(j)_{\textbf{updated}}$ and the corresponding new predicted pitting density $D(N_j)$. Subsequently, set $K_{\text{pitting}}(j+1) = K_{\text{pitting}}(j)_{\textbf{updated}}$ to predict the surface pitting density at time $j+1$.

Note that after attempts to update, the correlation coefficient between measured ICS2 and predicted surface pitting density cannot reach the 95% threshold, both $K_{\text{pitting}}(j)$ and $K_{\text{pitting}}(j+k)$ will be set to 0. This condition typically arises from a reduction in ICS2, whereas the proposed pitting model assumes ever-increasing pitting density. In practice, when surface pitting density starts to decrease, it means the gear should be replaced due to the presence of severe surface pitting [37]. This will be demonstrated in the results in Section 7.4.1.

7.4 TEST AND RESULTS

Two run-to-failure experiments were conducted under distinct lubrication conditions (lubricated and dry test) to assess the effectiveness of the proposed methodology in wear prediction.

As the details of test programs and experimental data collection have been presented in Chapter 3, they are not repeated here.

7.4.1 Monitoring and Prediction of Surface Pitting and Mild Tooth Profile Change During the Lubricated Test

Two wear events were presented in the lubricated test: surface pitting propagation and mild abrasive wear, leading to modest gear tooth profile changes. This enabled evaluation of the proposed method for both wear types using the same dataset. The collected moulds during the tests were imaged using a laser scanning confocal microscope (LSCM). The tooth profile change (see Figure 7.2 and Figure 7.3) and the actual pitting density (see Figure 7.4) were then quantified using the images and served as references to check the monitoring and prediction results [36]. Note that both the tooth profile change and pitting density in Figure 7.2, Figure 7.3, and Figure 7.4 are obtained from the pinion's mould images. The combined tooth profile change can be calculated based on the gear ratio, with the pinion's profile change accounting for 73% of the combined tooth profile change. As for surface pitting propagation, substantially less pitting occurred on the driven gear surface, and the occurrence of pitting on the driven gear was greatly delayed. Therefore, the total pitting density of the gear pair can be represented by the pinion's pitting density. The measured pitting density will be used to validate the effectiveness of the proposed method in surface pitting propagation monitoring and prediction. A run-in period took place from 0 to 0.12 million cycles, during which the surface asperities were worn away. Figure 7.4 shows that surface pitting started to propagate from 0.12 million cycles, and the pitting density mainly kept increasing until 2.12 million cycles. From 2.12 million cycles to 3.25 million cycles, the pitting density started to decrease. It appears that in this period, the generation of new pits was outweighed by the filling-in of existing cavities, likely due to wear debris being pressed into the cavities under meshing load [38].

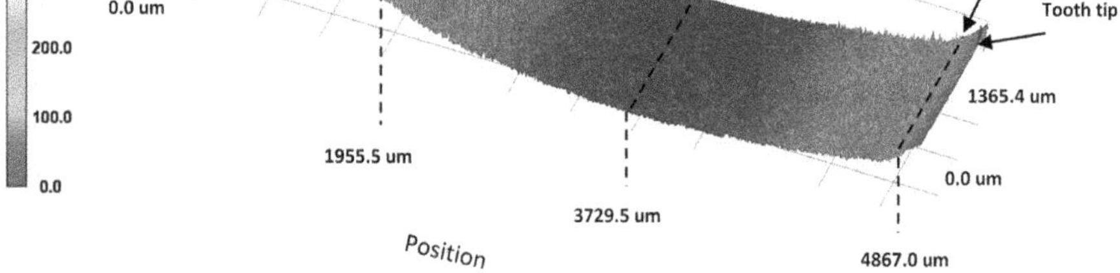

FIGURE 7.2 3D image of the mould profile collected from the pinion at 0.35 million cycles during the lubricated test. Made by the author [27].

7.4.1.1 Monitoring and Prediction of Change in Tooth Profile

This sub-section investigates the application and performance of the proposed methodology in monitoring and predicting wear-induced tooth profile changes. First, to determine the optimal vibration feature for tracking mild profile changes under lubricated conditions, the number of gear mesh harmonics included in RMS_{SA} (Eq. (7.1)) is evaluated. Then, with the help of the selected wear-relevant vibration feature, the profile change monitoring and prediction results are presented and compared with the actual measurements.

The continual lubrication provided by the oil bath resulted in a low wear rate and a relatively smooth worn tooth profile, as shown in Figure 7.3. Since gear teeth with different profile shapes can directly affect different gear mesh harmonics [39], this section examines how many harmonics are required to sufficiently represent the profile change. Figure 7.5 shows the change in gear mesh harmonics of measured vibrations throughout the test. It shows that the first two harmonics change more significantly than the others. To identify the optimal number of harmonics, a correlation analysis was performed between the vibration indicator RMS_{SA} and the wear depth (calculated based on mould images shown in Figure 7.3) with different numbers of gear mesh harmonics included (see Eq. (7.1)). More specifically, the Pearson correlation

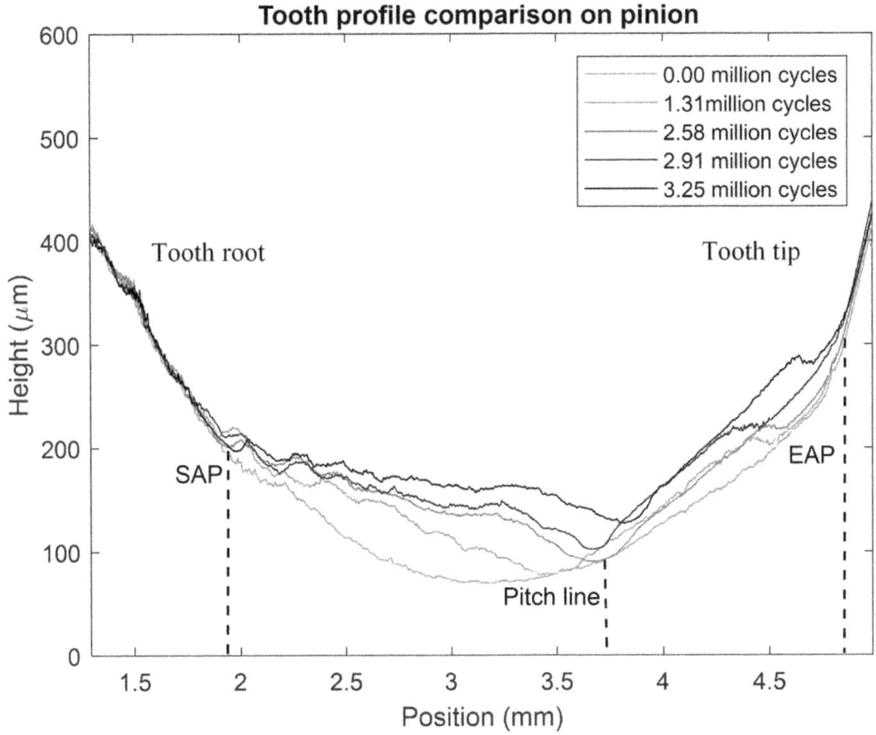

FIGURE 7.3 Lubricated test: The average profiles of the pinion at different running cycles. Made by the author [27].

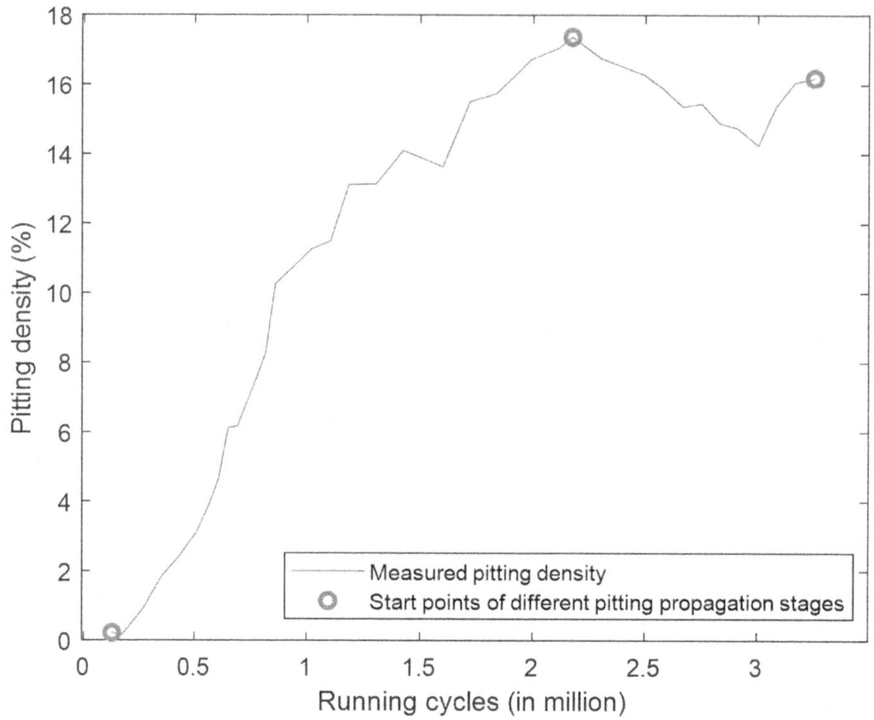

FIGURE 7.4 Lubricated test: Pitting density of the pinion obtained from moulds. Dots indicate 0.12 million cycles (pitting initiation), 2.12 million cycles (maximum pitting density), and 3.25 million cycles (test end). Made by the author [27].

coefficient is used to quantify the similarity between RMS_{SA} value and wear depth. The results indicate that tooth profile change can be represented with the first two gear mesh harmonics, and Figure 7.6 confirms a strong correlation between RMS_{SA} ($N = 2$) of measured vibration and wear depth. Consequently, in this book, RMS_{SA} ($N = 2$) is employed as the vibration feature to be compared with measured vibrations for the monitoring and prediction of tooth profile changes in the lubricated test.

For comparison purposes, the relationship between RMS_{SA} ($N = 2$) of measured vibration and measured pitting density is also presented in Figure 7.7. It reveals a poor relationship between RMS_{SA} ($N = 2$) and measured pitting density. Consequently, ICS2 is employed as the vibration feature in the proposed method for monitoring and predicting pitting propagation, as described later in this section

As described in Chapter 5, with the help of measured vibrations, the dynamic model was first calibrated at 0.12 million cycles (after the run-in period) to guarantee a sufficiently close match between simulated and experimental vibration signals and contact forces. Subsequently, the initial wear coefficient K_{wear} in the Archard wear model was calibrated to $3.55 \times 10^{-10}\,Pa^{-1}$ using the measured vibration signal at 0.27 million cycles. This calibration step constitutes the first update of the wear coefficient K_{wear}.

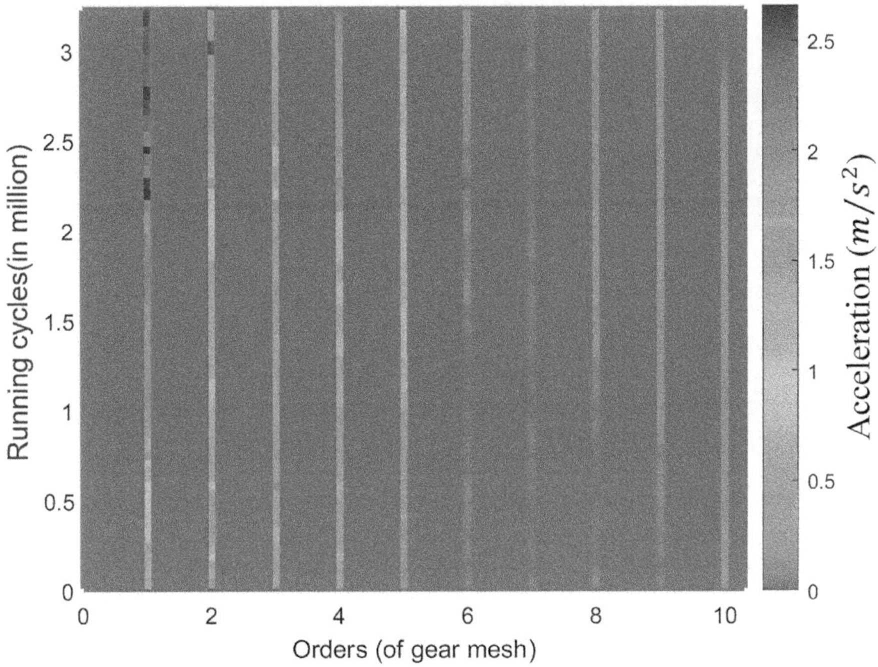

FIGURE 7.5 Gear mesh harmonics during the lubricated test. Made by the author [27].

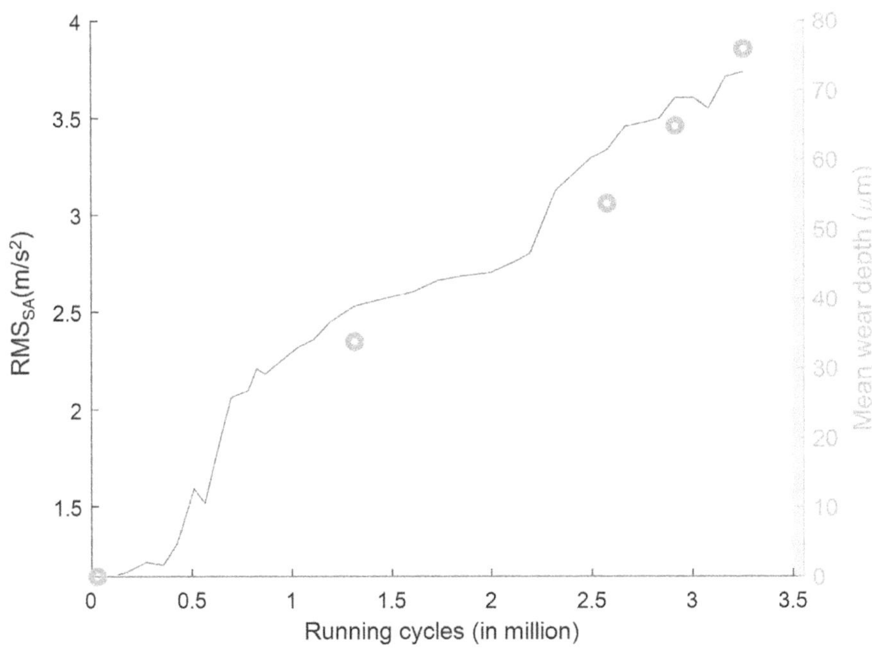

FIGURE 7.6 RMSSA(N=2)$^{RMS_{SA}}$ N = 2of measured vibrations and wear depth change from mould images in the lubricated test. Made by the author [27].

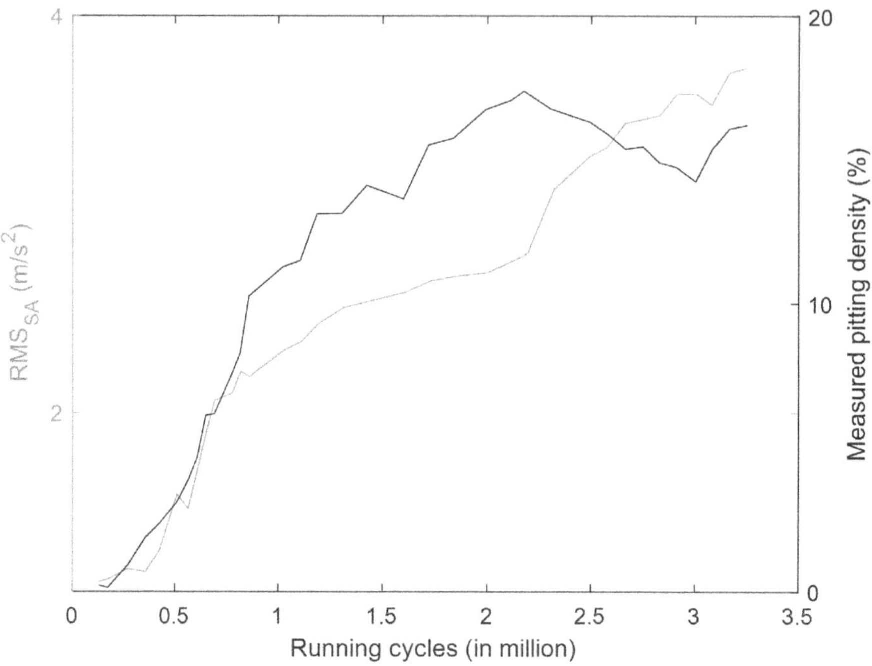

FIGURE 7.7 RMSSA(N=2)RMS$_{SA}$ N = 2 of measured vibrations and pitting density from the lubricated test. Made by the author.

In the further wear propagation process, because the error between the simulated and measured RMS$_{SA}$ values exceeds the predefined 5% threshold, three additional updates of K_{wear} were executed at 0.57, 1.43, and 2.19 million cycles. The updated wear coefficients are 4.03×10^{-10}, 1.44×10^{-10}, and 3.97×10^{-10} Pa^{-1}, respectively. The relationship between measured RMS$_{SA}$ and simulated RMS$_{SA}$ is given in Figure 7.8, showing that the error between them remains within 5.0% after necessary updating of the wear coefficient. The tooth profile change monitoring and prediction results are shown in Figure 7.9, illustrating that the maximum error between predicted and measured wear depth is 4.7%4.7 %. The specific points when updates were executed are marked by grey dashed lines in Figure 7.9. The tooth profile change is therefore well monitored and predicted using the proposed methodology.

Note that there are two wear rate changes observed in Figure 7.9. Possible explanations for these changes are as follows. The manufacturing marks on the driven gear have been significantly removed during [1.41~2.12] million cycles. The same phenomenon was observed in reference [36]. Thus, a plateau wear rate occurs during this period. After 2.12 million cycles, the involute profile of the gear tooth becomes flat, and the contact area of engaging tooth pairs increased. As a result, the tooth profile change rate might increase significantly [40]. Figure 7.9 also includes the monitoring and prediction results with several limited wear coefficient updates (not all necessary updates), specifically the first and third updates of K_{wear}. These results

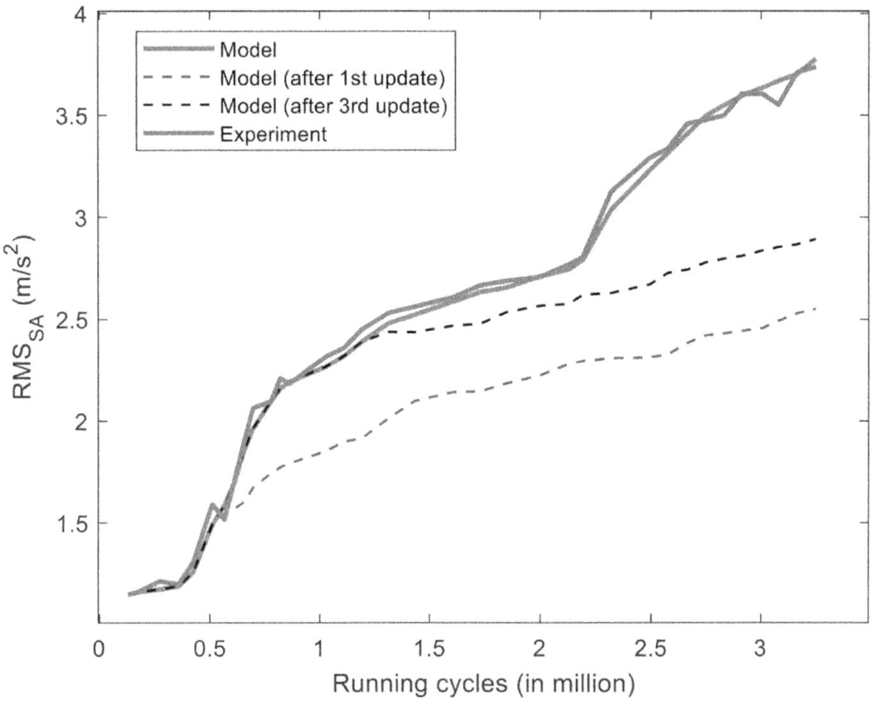

FIGURE 7.8 RMSSA(N=2)RMS$_{SA}$ comparison between experiment and model (lubricated test). Made by the author [27].

highlight the abrasive wear rate changed significantly during the lubricated test, and demonstrate that failing to update the wear coefficient would lead to substantial errors in wear depth prediction. In general, with more wear coefficient updates, increasingly accurate prediction results can be achieved. Thus, to ensure accurate wear prediction results, periodic comparisons with measurements and updates to the wear coefficient are recommended.

7.4.1.2 Surface Pitting Propagation Monitoring and Prediction

Observations from the lubricated test indicate that surface pitting initiates at the dedendum of the gear tooth, and then propagates to the pitch line and finally to the addendum, as shown in Figure 7.10. In the proposed methodology, with the dynamic contact force outputted from the dynamic model, the pitting propagation behaviour can be simulated using the developed surface pitting model, as shown in Figure 7.11. The results shown in Figure 7.11 indicate that, according to the developed model, the surface pitting also initiates at the gear tooth root and then propagates to the pitch line. The trends shown in Figure 7.11 match the actual pitting propagation trends shown in Figure 7.10, suggesting the developed surface pitting model has the ability to simulate accurately the surface pitting propagation behaviour. Note that the dynamic force exhibits several spikes, it is caused by the gear system resonances and

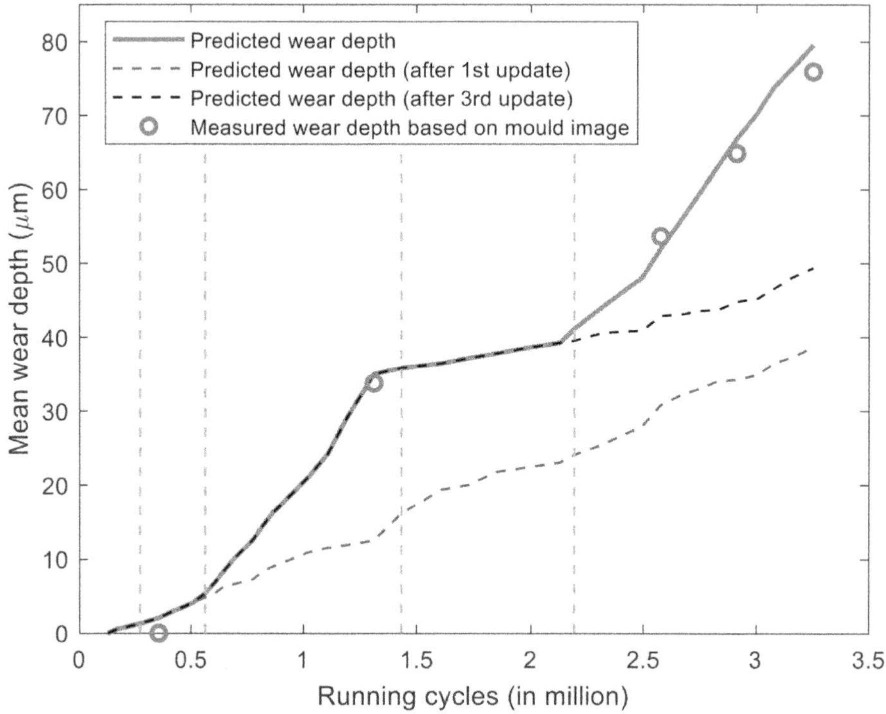

FIGURE 7.9 Mean wear depth comparison between experiment and model (lubricated test). Grey dashed lines show update points. Made by the author [27].

FIGURE 7.10 Optical microscope images (5× objective) of pinion tooth surfaces from the lubricated test. Image size: 2.8 mm × 2.11 mm. (a) 0.1296 M; (b) 0.8178 M; (c) 1.7175 M cycles. Photograph by the author [38].

a similar phenomenon was observed in reference [41]. More explanations of this phenomenon can refer to Section 5.3.

Using the developed surface pitting model and the proposed wear monitoring methodology, the surface pitting propagation can be monitored and predicted. Figure 7.12 shows the relationship between predicted pitting density and ICS2. It can be seen

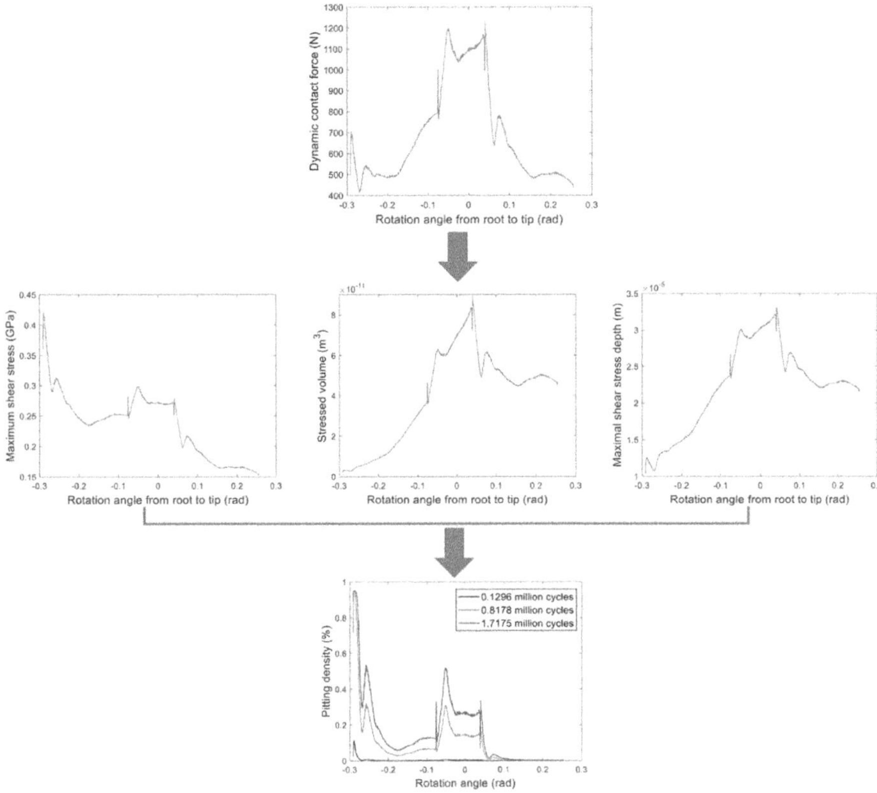

FIGURE 7.11 Pitting density calculation (surface pitting model). The pitch line at 0 rad is shown as a red continuous line. Made by the author [27].

that the ICS2 of measured vibration correlates well with the predicted pitting density, from 0.12 to 2.00 million cycles, following the run-in period. The magnitude of surface pitting coefficient K_{pitting} is initially calibrated as 5.64×10^{-15} using the mould images captured at 0.12 and 0.27 million cycles. This calibration of K_{pitting} represents the first update.

During the further pitting propagation process, six additional updates of $K_{pitting}$ were executed, prompted by Pearson's correlation coefficient C between measured ICS2 and predicted pitting density falling below the predefined 95%threshold. These updates occurred at 0.51, 0.56, 1.02, 1.11, 1.97, and 2.12 million cycles, respectively. The final update at 2.12 million cycles sets K_{pitting} to 0 because the 95% threshold for C cannot be achieved. Figure 7.13 shows the pitting prediction results compared with the actual measured pitting densities using mould images. The comparison results demonstrate that the surface pitting propagation is very well monitored and predicted following all necessary updates prior to 2.00 million cycles.

Note that from 2.00 to 3.25 million cycles, the measured pitting density decreased slightly. The tooth profile started to change rapidly in this period due to intensified

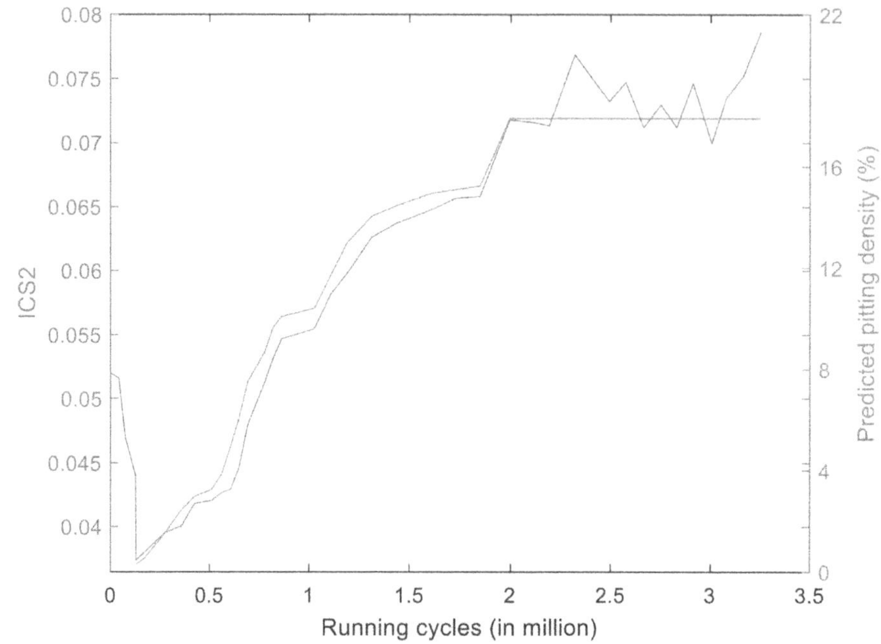

FIGURE 7.12 ICS2 and predicted pitting density comparison (updated model). Made by the author [27].

abrasive wear. Because the developed surface pitting model is designed to represent accumulated pitting behaviour on the assumption that pitting density cannot decrease, the stabilisation stage (2.00 million cycles to 3.25 million cycles) was not simulated and predicted in this book. Consequently, the pitting coefficient is set to 0 automatically during this interval, and the predicted surface pitting density remains a constant value. Furthermore, at 2.00 million cycles, the pitting density is already 17.4%, and according to the ASM Handbook [37], the gear should be scrapped. Thus, the gear is considered to have already failed. The proposed methodology therefore demonstrates excellent pitting prediction results over the most meaningful phase of the gear's degradation, from 0.12 to 2.00 million cycles. Figure 7.13 shows the prediction results obtained using only the earlier updates of the pitting coefficient, the first and the fourth. The results indicate that the 'point of failure' (~2 million cycles) is in fact better predicted using the fourth update of K_{pitting} than the first, and a more accurate prediction result can be expected with more updates.

Note that in the proposed methodology, the tooth profile is continually updated through the improved Archard wear model (introduced in Chapter 6), providing updated contact pressure to the surface pitting model for pitting propagation prediction. Conversely, the occurrence of surface pitting affects the effective contact width, resulting in a change in the contact pressure. This is a two-way relationship between surface pitting propagation and tooth profile change. The comparison analysis between measurements and simulations can guarantee the accuracy of the predictions

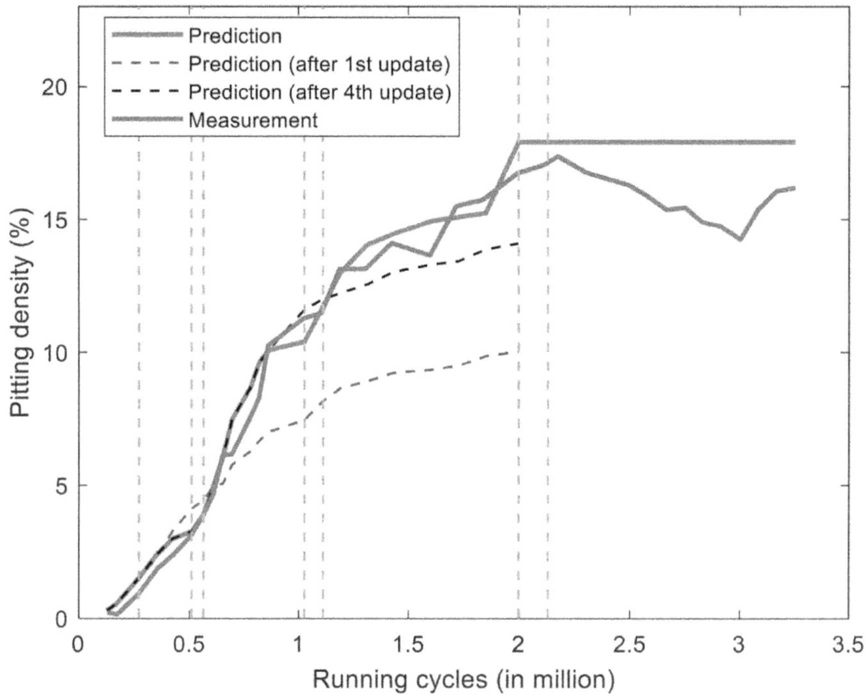

FIGURE 7.13 Pitting density comparison between experiment and model-based prediction. Grey dashed lines show update points. Made by the author [27].

for both tooth profile change and surface pitting propagation, as shown in Figure 7.9 and Figure 7.13.

7.4.2 Monitoring and Prediction of Severe Tooth Profile Change During the Dry Test

Compared with the lubricated test, the dry test represents an extreme scenario characterised by significant abrasive wear and a correspondingly large tooth profile change, while very little surface pitting propagation (as shown in Figure 7.14). The proposed approach is however still effective under dry conditions, as will be demonstrated in this section.

As with the lubricated test, this section first investigates the number of gear mesh harmonics needed in the vibration feature to accurately track the severe tooth profile changes. Figure 7.15 shows the amplitude evolution of the first ten gear mesh harmonics from the measured acceleration signal during the test, indicating that the first six harmonics change significantly. To establish the required number of harmonics for tracking profile changes in this test, correlation analysis was performed between the wear depth (calculated based on the weight of collected wear particles) and the vibration indicator RMS_{SA} with different numbers of gear mesh harmonics (absolute

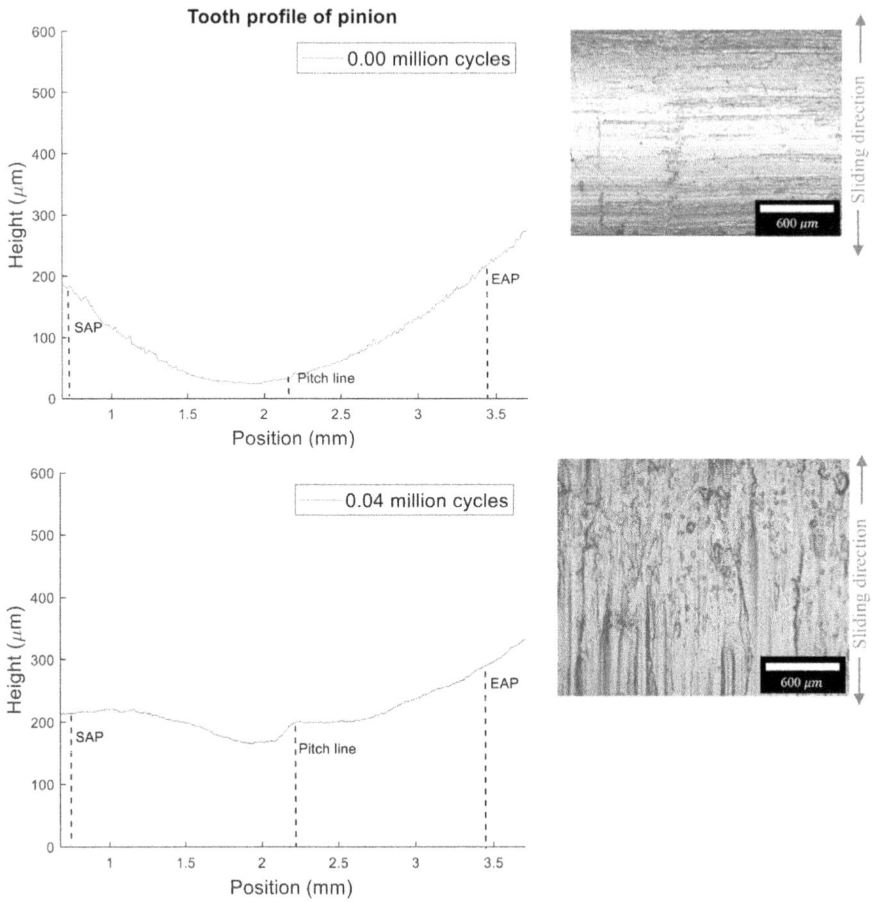

FIGURE 7.14 Gear tooth profile and surface morphology of the dry test at 0 and 0.04 million cycles (SAP: start of active profile, EAP: end of active profile). Made by the author [27].

values). The results demonstrate that severe tooth profile changes can be represented with the first six-gear mesh harmonics. Figure 7.16 confirms this by showing the strong relationship between measured RMS_{SA} (N = 6) and measured wear depth based on collected wear particles. Consequently, for this test, RMS_{SA} (N = 6) served as the vibration feature to compare the measured and simulated vibrations for tooth profile change monitoring and prediction. Compared with the lubricated test, severe tooth profile change occurs during the dry test; therefore, more gear mesh harmonics (the first six) are required to represent the worn tooth profile, while only the first two gear mesh harmonics are required to represent the tooth profile change during lubricated tests. This phenomenon has also been discussed in Section 7.2.

The Archard wear model parameter K_{wear} was updated when necessary with the procedure discussed in Section 7.3.3 at each measured time. Figure 7.17 and

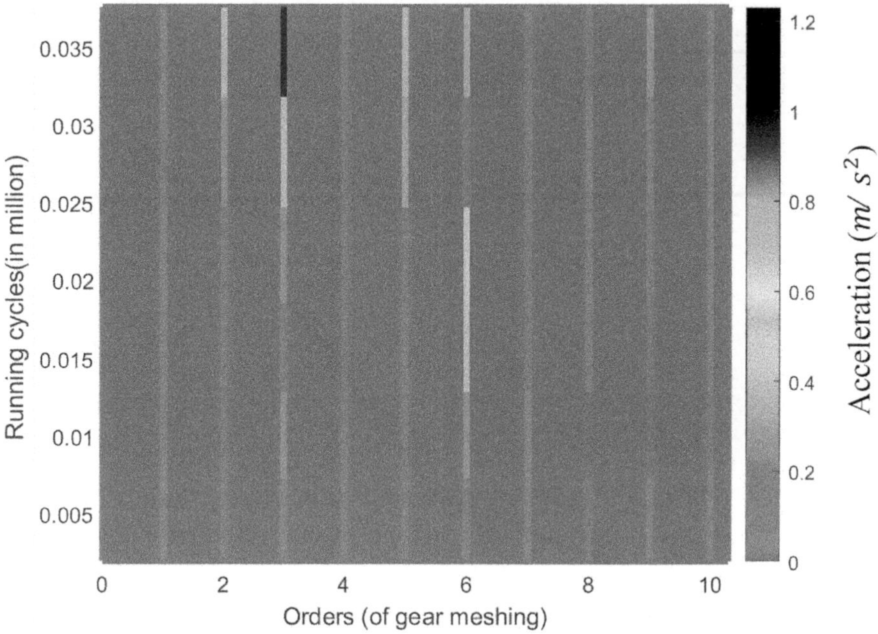

FIGURE 7.15 Changes in the first ten gear mesh harmonics during the dry test. Made by the author [27].

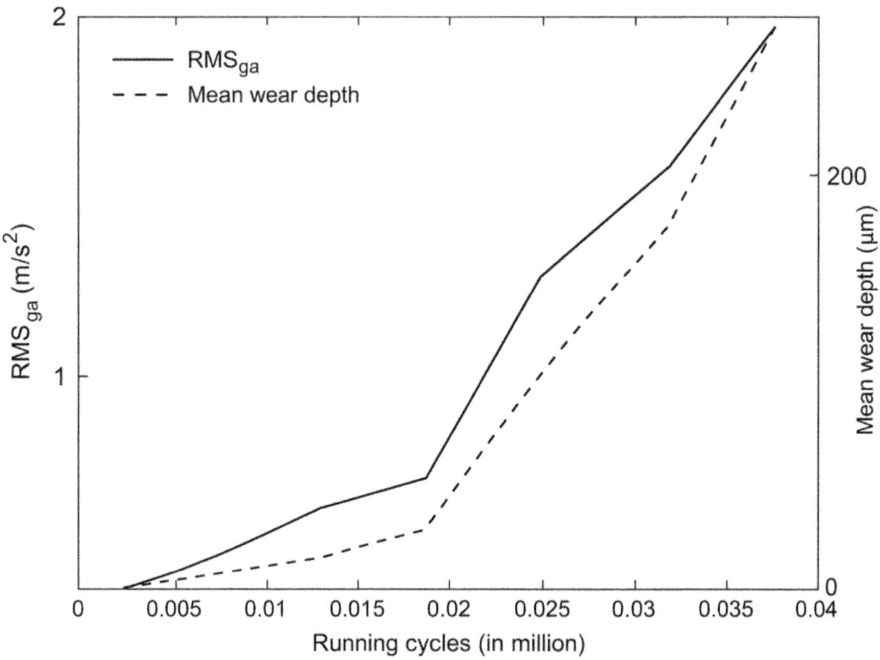

FIGURE 7.16 Dry test: measured RMS_{SA}(N = 6) and wear depth based on collected wear particles. Made by the author [27].

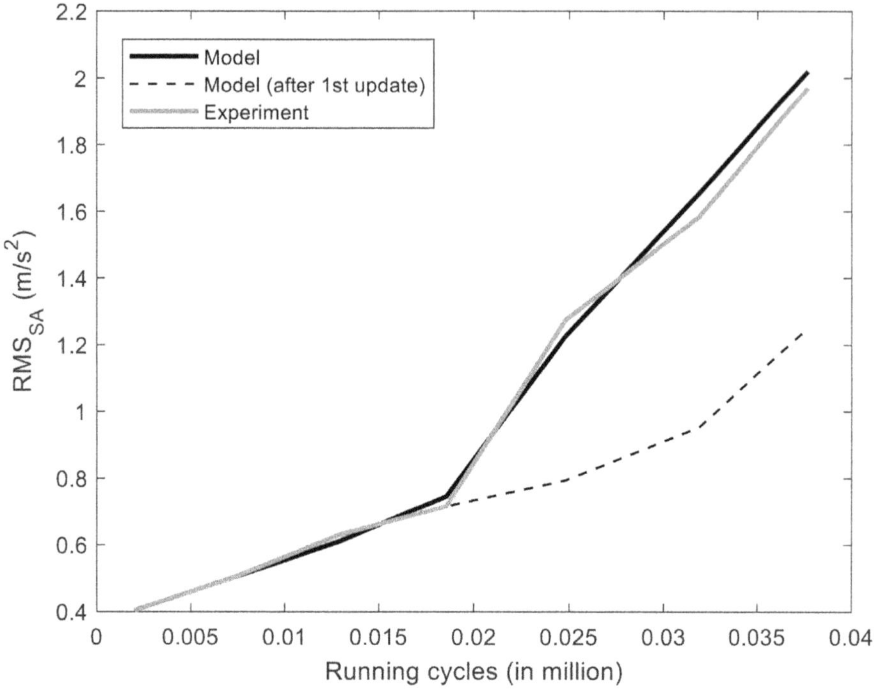

FIGURE 7.17 RMS_{SA} comparison results of the dry test: Experiment and model (after updating). Made by the author [27].

Figure 7.18, respectively, show the trends of RMS_{SA} and mean wear depth versus wear cycles (pinion rotations). Two updates were executed during the short dry test, when the error between the RMS_{SA} value of the simulated and measured vibration signals exceeded the predefined 5% threshold. Using measured vibration at 0.007 million running cycles, the wear model parameter was found/calibrated to be $1.65\times10^{-6}\,Pa^{-1}$ and the predicted mean wear depth was estimated at $12.92\mu m$. This calibration constituted the first update. Subsequently, the simulated RMS_{SA} and estimated wear depth match well with experiments until the second update was triggered at 0.025 million cycles, where the wear model parameter K_{wear} was updated to $2.19\times10^{-6}\,Pa^{-1}$. Following this update, the wear model provided accurate prediction results for the remainder of the test. Figure 7.17 indicates that the maximum error between predicted wear depth and measured wear depth is 4.5%. Therefore, with a vibration feature based on gear mesh harmonics combined with an improved Archard model, the proposed surface degradation monitoring and prediction methodology has an excellent performance in tracking and predicting the tooth profile change (in terms of wear depth) under dry conditions.

The tooth profile prediction results with one update/calibration of K_{wear} are also shown in Figure 7.18 for comparison purposes. The results reveal that the wear rate

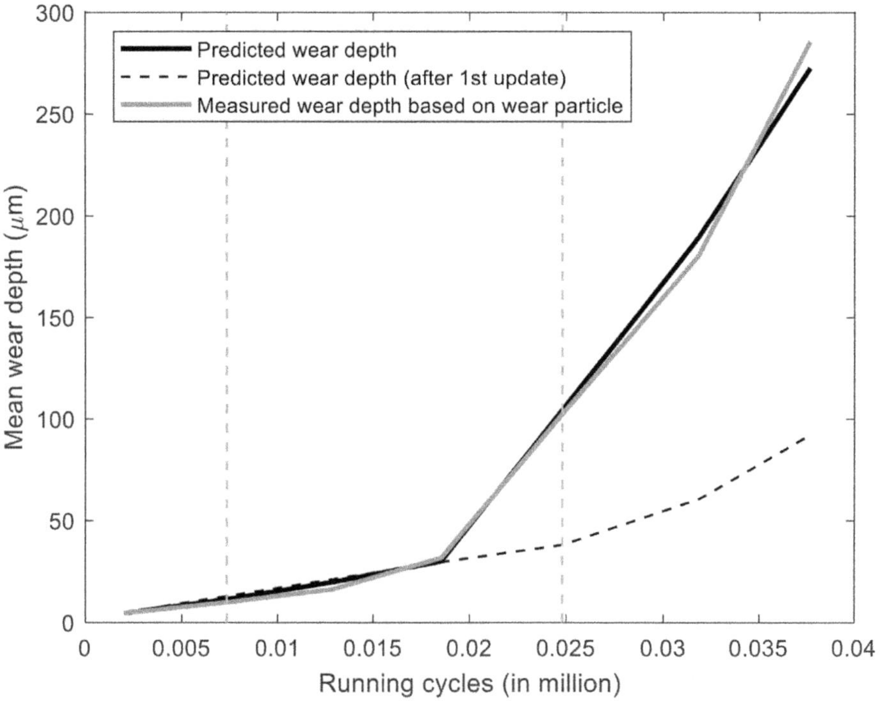

FIGURE 7.18 Mean wear depth comparison between experiment and model (dry test). Grey dashed lines show update points. Made by the author [27].

changes significantly around 0.018 million cycles. Without the second updating of K_{wear}, there would be a significant under-estimation of the future wear level.

The dry test exhibited negligible surface pitting from the mould images (as shown in Figure 7.14), while the ICS2 of the measured vibrations exhibited fluctuation during the wear propagation, as shown in Figure 7.19. Based on the rules introduced in Section 7.3.3, the surface pitting coefficient was set to zero, and the predicted pitting density was 0% during the dry test. This is consistent with the observation in reference [42] that the dominant wear phenomenon in this test is tooth profile change, with very little or no surface pitting propagation.

7.5 SUMMARY

This chapter presents a vibration-based surface degradation monitoring and prediction methodology for monitoring and predicting the two common wear phenomena in gears: Gear tooth profile change and surface pitting propagation. Based on the published research and discussion in Section 7.1, there is no digital twin approach that can monitor and predict the two wear phenomena simultaneously. In the proposed methodology, a simple vibration feature based on a few gear mesh harmonics is used for tracking gear tooth profile changes under different lubrication conditions, and an

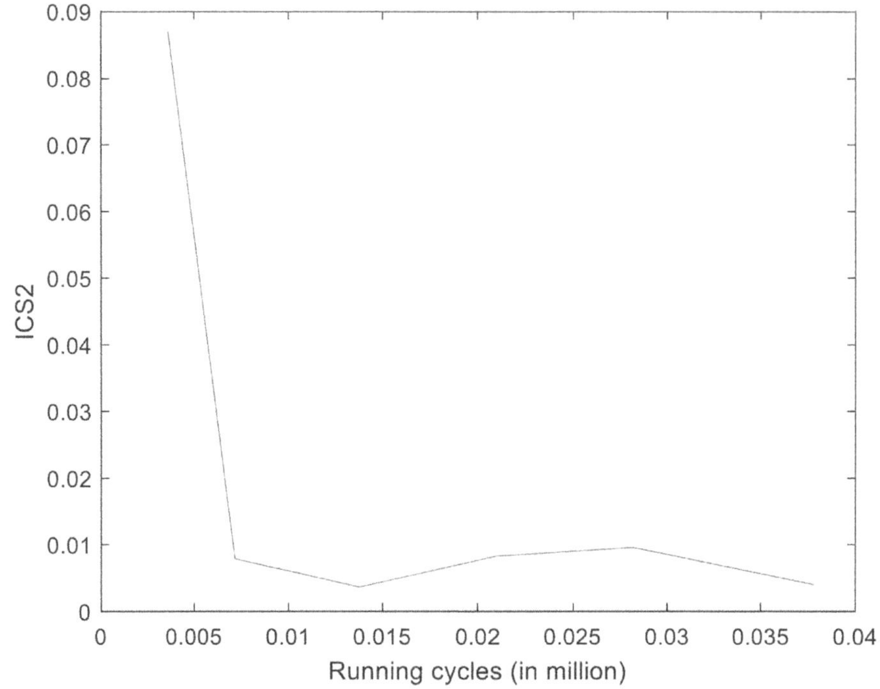

FIGURE 7.19 Dry test: ICS2 of measured vibrations. Made by the author [27].

enhanced Archard wear model (introduced in Chapter 6) with a gearbox dynamic model is used to predict future worn tooth profiles. An effective and efficient surface pitting model was also proposed in this chapter, employing the level of second-order cyclostationarity (ICS2) from the measured vibration signal for tracking surface pitting progression.

A key component of the methodology is that through regular and intelligent use of measured vibration signals, the parameters of the two wear models can be updated as necessary, thereby progressively ensuring wear prediction accuracy. Unique among existing studies in gear-wear prediction, the interaction of the developed dynamic and wear models, along with this updating capability, delivers an approach that is able to monitor and predict both profile change and surface pitting simultaneously under different lubrication conditions. Although two calibration steps are required by the approach – one for the dynamic model and one for relating ICS2 to the pitting level – the models are relatively simple. The flexible and evolving nature of the prediction approach means that it could be easily deployed within existing digital twin frameworks, bringing significant potential benefits to gear prognostics in practice. In summary, this chapter introduces novel gear surface degradation prediction models and schemes. Through regular and intelligent use of measured vibration signals, the models can be updated as necessary, ensuring accurate predictions of gear wear propagation can be delivered.

The primary limitation of the method in its current form is that a scan (or perhaps a close visual inspection) of the gear surface is required to find the 'scale factor' between ICS2 and the actual pitting level. This necessitates the gearbox stoppage and partial dismantling. Another option is that the scan could presumably be done by endoscopic examination, with almost no dismantling, just access through a small inspection cover; this is widely practised in wind turbine gearboxes [43]. It has also been found that ICS2 tends not to track pitting levels well in the initial stages. Consequently, to deliver improved pitting propagation predictions, a valuable direction for future work is to establish a more definitive relationship between ICS2 and pitting level or to develop a more reliable pitting-sensitive vibration indicator.

REFERENCES

[1] C. Hu, W.A. Smith, R.B. Randall, Z. Peng, Development of a gear vibration indicator and its application in gear wear monitoring, *Mechanical Systems and Signal Processing*, 76–77 (2016) 319–336. https://doi.org/10.1016/j.ymssp.2016.01.018.

[2] Y. Pan, R. Hong, J. Chen, J. Singh, X. Jia, Performance degradation assessment of a wind turbine gearbox based on multi-sensor data fusion, *Mechanism and Machine Theory*, 137 (2019) 509–526. https://doi.org/10.1016/j.mechmachtheory.2019.03.036.

[3] Y. Pan, R. Hong, J. Chen, W. Wu, A hybrid DBN-SOM-PF-based prognostic approach of remaining useful life for wind turbine gearbox, *Renewable Energy*, 152 (2020) 138–154. https://doi.org/10.1016/j.renene.2020.01.042.

[4] J.P. Salameh, S. Cauet, E. Etien, A. Sakout, L. Rambault, Gearbox condition monitoring in wind turbines: A review, *Mechanical Systems and Signal Processing*, 111 (2018) 251–264. https://doi.org/10.1016/j.ymssp.2018.03.052.

[5] M. Weibring, L. Gondecki, P. Tenberge, Simulation of fatigue failure on tooth flanks in consideration of pitting initiation and growth, *Tribology International*, 131 (2019) 299–307. https://doi.org/10.1016/j.triboint.2018.10.029.

[6] S.M. Evans, P.S. Keogh, Wear mechanisms in polyoxymethylene spur gears, *Wear*, 428–429 (2019) 356–365. https://doi.org/10.1016/j.wear.2019.03.027.

[7] A.C. Petare, N. Kumar Jain, On simultaneous improvement of wear characteristics, surface finish and microgeometry of straight bevel gears by abrasive flow finishing process, *Wear*, 404–405 (2018) 38–49. https://doi.org/10.1016/j.wear.2018.03.002.

[8] R.I. Trezona, D.N. Allsopp, I.M. Hutchings, Transitions between two-body and three-body abrasive wear: Influence of test conditions in the microscale abrasive wear test, *Wear*, 225–229 (1999) 205–214. https://doi.org/10.1016/S0043-1648(98)00358-5.

[9] H. Ding, A. Kahraman, Interactions between nonlinear spur gear dynamics and surface wear, *Journal of Sound and Vibration*, 307 (2007) 662–679. https://doi.org/10.1016/j.jsv.2007.06.030.

[10] J. Archard, Contact and rubbing of flat surfaces, *Journal of Applied Physics*, 24 (1953) 981–988.

[11] D. Huang, Z. Wang, A. Kubo, Hypoid gear integrated wear model and experimental verification design and test, *International Journal of Mechanical Sciences*, 166 (2020) 105228. https://doi.org/10.1016/j.ijmecsci.2019.105228.

[12] K. Aslantaş, S. Taşgetiren, A study of spur gear pitting formation and life prediction, *Wear*, 257 (2004) 1167–1175. https://doi.org/10.1016/j.wear.2004.08.005.

[13] C.H. Chue, H.H. Chung, Pitting formation under rolling contact, *Theoretical and Applied Fracture Mechanics*, 34 (2000) 1–9. https://doi.org/10.1016/S0167-8442(00)00019-7.
[14] S. Li, A. Kahraman, A fatigue model for contacts under mixed elastohydrodynamic lubrication condition, *International Journal of Fatigue*, 33 (2011) 427–436. https://doi.org/10.1016/j.ijfatigue.2010.09.021.
[15] G.E. Morales-Espejel, P. Rycerz, A. Kadiric, Prediction of micropitting damage in gear teeth contacts considering the concurrent effects of surface fatigue and mild wear, *Wear*, 398–399 (2018) 99–115. https://doi.org/10.1016/j.wear.2017.11.016.
[16] E. Charkaluk, A. Constantinescu, H. Maïtournam, K. Dang Van, Revisiting the Dang Van criterion, *Procedia Engineering*, 1 (2009) 143–146. https://doi.org/10.1016/j.proeng.2009.06.033.
[17] E. Charkaluk, A. Bignonnet, A. Constantinescu, K. Dang Van, Fatigue design of structures under thermomechanical loadings, *Fatigue & Fracture of Engineering Materials & Structures*, 25 (2002) 1199–1206. https://doi.org/10.1046/j.1460-2695.2002.00612.x.
[18] K.D. Van, M.H. Maitournam, On some recent trends in modelling of contact fatigue and wear in rail, *Wear*, 253 (2002) 219–227. https://doi.org/10.1016/S0043-1648(02)00104-7.
[19] G. Lundberg, Dynamic capacity of rolling bearings, *IVA Handlingar*, 196 (1947) 12.
[20] B. Jalalahmadi, F. Sadeghi, V. Bakolas, Material Inclusion Factors for Lundberg-Palmgren–based RCF life equations, *Tribology Transactions*, 54 (2011) 457–469. https://doi.org/10.1080/10402004.2011.560412.
[21] T.E. Tallian, Simplified contact fatigue life prediction model – Part I: Review of published models, *Journal of Tribology*, 114 (1992) 207–213. https://doi.org/10.1115/1.2920875.
[22] E.V. Zaretsky, J.V. Poplawski, S.M. Peters, Comparison of life theories for rolling-element bearings, *Tribology Transactions*, 39 (1996) 237–248. https://doi.org/10.1080/10402009608983525.
[23] American Gear Manufacturers Association, *Appearance of gear teeth: Terminology of wear and failure*, Vol 40, American Gear Manufacturers Association, Virginia, United States, 1995.
[24] R.B. Randall, A new method of modeling gear faults, *Journal of Mechanical Design*, 104 (1982) 259–267. https://doi.org/10.1115/1.3256334.
[25] A. Raad, J. Antoni, M. Sidahmed, Indicators of cyclostationarity: Theory and application to gear fault monitoring, *Mechanical Systems and Signal Processing*, 22 (2008) 574–587. https://doi.org/10.1016/j.ymssp.2007.09.011.
[26] W.A. Smith, R.B. Randall, X.d.C. du Mée, P. Peng, *Use of cyclostationary properties to diagnose planet bearing faults in variable speed conditions*, 10th DST group international conference on health and usage monitoring systems, 17th Australian Aerospace Congress, Engineers Australia, Royal Aeronautical Society, 2017, pp. 26–28.
[27] K. Feng, W.A. Smith, R.B. Randall, H Wu, Z. Peng, Vibration-based monitoring and prediction of surface profile change and pitting density in a spur gear wear process, *Mechanical Systems and Signal Processing*, 165 (2022) 108319. https://doi.org/10.1016/j.ymssp.2021.108319
[28] H. Liu, H. Liu, C. Zhu, J. Tang, Study on gear contact fatigue failure competition mechanism considering tooth wear evolution, *Tribology International*, 147 (2020) 106277. https://doi.org/10.1016/j.triboint.2020.106277.

[29] J.J. Coy, D.P. Townsend, E.V. Zaretsky, *An update on the life analysis of spur gears*, NASA Technical Report, NASA Technical Reports Server, 1983.

[30] D. Zhu, N. Ren, Q.J. Wang, Pitting life prediction based on a 3D line contact mixed EHL analysis and subsurface von mises stress calculation, *Journal of Tribology*, 131 (2009) 041501. https://doi.org/10.1115/1.3195040.

[31] G.E. Totten, *ASM handbook, Volume 18: Friction, lubrication, and wear technology*, ASM international, Cleveland, 1992.

[32] T.A. Harris, M.N. Kotzalas, *Rolling bearing analysis: Essential concepts of bearing technology*, CRC Press, Boca Raton, United States, 2007.

[33] E. Ioannides, T.A. Harris, A new fatigue life model for rolling bearings, *Journal of Tribology*, 107 (1985) 367–377. https://doi.org/10.1115/1.3261081.

[34] A.G. Asuero, A. Sayago, A. Gonzalez, The correlation coefficient: An overview, *Critical Reviews in Analytical Chemistry*, 36 (2006) 41–59. https://doi.org/10.1080/10408340500526766.

[35] P. Sedgwick, Pearson's correlation coefficient, *BMJ*, 345 (2012) e4483–e4483. https://doi.org/10.1136/bmj.e4483.

[36] H. Chang, P. Borghesani, W.A. Smith, Z. Peng, Application of surface replication combined with image analysis to investigate wear evolution on gear teeth – A case study, *Wear*, 430–431 (2019) 355–368. https://doi.org/10.1016/j.wear.2019.05.024.

[37] F.C. Campbell, *Fatigue and fracture: Understanding the basics*, ASM International, Almere, Netherlands, 2012.

[38] P. Feng, P. Borghesani, H. Chang, W.A. Smith, R.B. Randall, Z. Peng, Monitoring gear surface degradation using cyclostationarity of acoustic emission, *Mechanical Systems and Signal Processing*, 131 (2019) 199–221. https://doi.org/10.1016/j.ymssp.2019.05.055.

[39] W.D. Mark, *Performance-based gear metrology*, Wiley Online Library, 2012.

[40] H. Chang, P. Borghesani, Z. Peng, Investigation on the relationship between macropits and wear particles in a gear fatigue process, *Wear*, 484–485 (2021) 203724. https://doi.org/10.1016/j.wear.2021.203724.

[41] H. Dai, F. Chen, C. Xun, X. Long, Numerical calculation and experimental measurement for gear mesh force of planetary gear transmissions, *Mechanical Systems and Signal Processing*, 162 (2022) 108085. https://doi.org/10.1016/j.ymssp.2021.108085.

[42] S. Dhanasekaran, R. Gnanamoorthy, Gear tooth wear in sintered spur gears under dry running conditions, *Wear*, 265 (2008) 81–87. https://doi.org/10.1016/j.wear.2007.08.025.

[43] R. Zimroz, W. Bartelmus, T. Barszcz, J. Urbanek, Diagnostics of bearings in presence of strong operating conditions non-stationarity – A procedure of load-dependent features processing with application to wind turbine bearings, *Mechanical Systems and Signal Processing*, 46 (2014) 16–27. https://doi.org/10.1016/j.ymssp.2013.09.010.

8 Conclusions and Future Directions

8.1 SUMMARY OF FINDINGS AND CONTRIBUTION TO RESEARCH

This book investigated the interconnection between vibration characteristics and gear wear through signal processing algorithms and modelling techniques to enable accurate gear wear monitoring and prediction using vibration-based techniques. This section summarises the main outcomes of this book, guided by research objectives.

- **Identification of gear wear mechanism and tracking wear evolution using cyclostationary properties of measured vibrations (Objective 1).**

A novel relationship is proposed in Chapter 4 to describe the internal connection between vibration characteristics and gear surface features. Specifically, the link between the carrier frequency of vibrations f_v and spatial frequency of gear surface f_s was built. The relationship is validated through the cyclostationarity analysis of vibrations and the power spectral density (PSD) analysis of the scanned gear surface images.

Based on the understanding gained from the derived equation, a vibration-based gear wear mechanism identification approach is proposed. Specifically, through the use of cyclostationary properties of vibrations, the abrasive wear and fatigue pitting are separated and identified via spectral coherence map and ICS2-based band selection results. This developed novel online gear wear mechanism identification approach avoids disrupting gearbox operation (in contrast to visual inspection, the most widely used approach for identification of wear mechanism at present) or delaying the analysis of wear debris generated from gear surface, making it a more practical tool in industry practice.

Furthermore, with the help of the derived equation and vibration cyclosationarity indicator, an informative vibration frequency band can be determined, then the fatigue pitting and abrasive wear propagation can be well tracked and monitored using ICS2 of vibration in the selected appropriate frequency bands. Different from previous works, the carrier frequencies (spectral content) of the gearmesh-cyclic CS2 components are analysed and used in this book to distinguish and track the two wear phenomena.

 DOI: 10.1201/9781003686736-8

- **Dynamic model development (Objective 2).**

A comprehensive dynamic model is developed in Chapter 5, based on the UNSW spur gearbox test rig, and there is no published work on modelling this test rig, whose layout is shown in Figure 3.2. Following a series of necessary model validations and calibrations, the natural frequencies and time waveforms of the developed dynamic model match well with the measurements from the actual gearbox transmission system. As a result, the dynamic model can generate contact force and vibration responses, which are close enough to the actual test rig. The developed dynamic model can provide insights into the coupling effects between gear wear and gear dynamic characteristics (e.g. contact force and vibrations). This can benefit the gear maintenance schedule to minimise the consequences of gear wear on gearbox service life.

- **Monitoring and prediction of tooth profile changes during wear progression (Objective 3).**

A novel vibration-based updating scheme is developed in Chapter 6 to monitor and predict the gear tooth profile change during gear wear propagation. Unique to previously published work, such as references [1–3], the proposed scheme compares measured vibrations with simulated vibrations from the dynamic model, to update the coefficient K_{wear} when a deviation from predictions is detected. The developed methodology can track and correct for changes in the gear wear rates, thereby enabling reliable gear wear prediction. In addition, the vibration signals can be easily acquired without disturbing gearbox operation, and no gearbox stoppage or disassembly is required to obtain the wear mass, as presented in reference [4].

In the proposed wear prediction scheme, the Archard wear model is improved with consideration of the effect of Hertzian deformation, giving a contact area rather than a line. The improvement makes the calculated wear profile distribution more reasonable and realistic compared with the original Archard wear model used in the study [1].

- **Development of a digital twin approach for monitoring and prediction of surface pitting and tooth profile changes (Objective 4).**

A vibration-based methodology is proposed in Chapter 7 for monitoring and predicting the two common wear phenomena in gears: Gear profile change (e.g. from abrasive wear), and surface pitting propagation. Through regular and intelligent use of measured vibration signals, the parameters of the wear models can be updated as necessary, ensuring wear predictions that tend to improve over time. It is a novel work since no research has reported a digital twin methodology that can predict two wear events' propagation and validate its effectiveness using measurements from gearbox test rig.

To simulate the surface pitting propagation behaviours, an efficient pitting propagation model is also developed in Chapter 7, based on the Lundberg Palmgren model

[5]. With the developed fatigue pitting model, the accumulated pitting propagation process can be simulated, and the impacts on pitting propagation rate from the prior pitting are also taken into consideration.

To conclude, this book has comprehensively investigated the wear propagation phenomenon and its consequences on gear tooth surface with the use of vibration analysis techniques, including signal processing algorithms and modelling methodology. To achieve gear wear monitoring, a vibration-based integrated system is developed. With the developed integrated system, the gear wear mechanism can be identified online, and then its propagation can be reliably tracked through the proposed digital twin approach, so that the remaining useful life of the gear system can be predicted. The developments for gear wear monitoring and prediction in this book bring significant benefits to the research community and industrial practices.

8.2 RECOMMENDATIONS FOR FUTURE WORK

In this book, vibration-based gear wear monitoring techniques are developed and applied to wear mechanism identification and wear prediction of the fixed-axis gearbox. Considering that gearbox transmission systems share similar kinematic characteristics, the techniques developed in this book should be tested on other gear types and gearing arrangements, including planetary gearboxes, and industrial practices.

Even though the theoretical developments in this book are validated through the relevant experimental investigations, some limitations still exist in the proposed techniques, which will be pointed out as follows, along with corresponding recommendations for future improvements.

1. the limitation of using ICS2 to monitor the fatigue pitting and abrasive wear propagation is that the actual wear severity cannot be directly assessed from the ICS2 trend, since the mathematical relationship between ICS2 trends and the real severity of gear wear has not been established in Chapter 4. Therefore, future work will focus on establishing this mathematical relationship using the regression analysis theory [6, 7], enabling the real wear severity to be quantified once the measured vibration is acquired.
2. a further limitation of using the Archard wear model to represent the wear-induced tooth profile change. The Archard model captures the significant component of the geographical variation of wear along the tooth flank; more complex dynamics can contribute significantly to the evolution of the wear profile geometry. Consequently, future work will consider model development that has the capability of modelling complex wear profile geometry, such as using Legendre Polynomials [8].
3. in the vibration-based gear wear monitoring and prediction methodology developed in Chapter 7, the ICS2 of measured vibration is compared with the predicted fatigue pitting density, and then update the surface pitting coefficient $K_{pitting}$ if necessary. This approach should deliver an accurate prediction of overall surface pitting severity representing an average across the whole

surface; however, it cannot provide information on the pitting distribution along the gear tooth (from root to tip). Therefore, vibration features that can accurately represent the surface pitting distribution are of significant value and should be explored in the future.

REFERENCES

[1] H. Ding, A. Kahraman, Interactions between nonlinear spur gear dynamics and surface wear, *Journal of Sound and Vibration*, 307 (2007) 662–679. https://doi.org/10.1016/j.jsv.2007.06.030.

[2] D. Park, M. Kolivand, A. Kahraman, Prediction of surface wear of hypoid gears using a semi-analytical contact model, *Mechanism and Machine Theory,* 52 (2012) 180–194. https://doi.org/10.1016/j.mechmachtheory.2012.01.019.

[3] Z. Zhang, C. Zheng, M. Wen, S. Yang, H. Li , Q. Du, *A wear prediction model for spur gears based on the dynamic meshing force and tooth profile reconstruction*, 2nd International Conference on Computer Engineering, Information Science & Application Technology (ICCIA 2017), Atlantis Press, 2016.

[4] F. Zhao, Z. Tian, X. Liang, M. Xie, An integrated prognostics method for failure time prediction of gears subject to the surface wear failure mode, *IEEE Transactions on Reliability*, 67 (2018) 316–327. https://doi.org/10.1109/TR.2017.2781147.

[5] G. Lundberg, Dynamic capacity of rolling bearings, *IVA Handlingar*, 196 (1947) 12.

[6] S. Chatterjee, A.S. Hadi, *Regression analysis by example*, Wiley, Hoboken, New Jersey, United States, 2015.

[7] E. Uludamar, E. Tosun, K. Aydın, Experimental and regression analysis of noise and vibration of a compression ignition engine fuelled with various biodiesels, *Fuel*, 177 (2016) 326–333. https://doi.org/10.1016/j.fuel.2016.03.028.

[8] W.D. Mark, *Performance-based gear metrology*, Wiley Online Library, Chichester, United Kingdom, 2012.

Index

Z

For Product Safety Concerns and Information please contact our EU
representative GPSR@taylorandfrancis.com
Taylor & Francis Verlag GmbH, Kaufingerstraße 24, 80331 München, Germany

www.ingramcontent.com/pod-product-compliance
Ingram Content Group UK Ltd.
Pitfield, Milton Keynes, MK11 3LW, UK
UKHW020206200926
475702UK00004B/633

* 9 7 8 1 0 4 1 1 6 7 2 4 2 *